PHILADELPHIA'S STREETCAR HERITAGE

Looking west on Market Street from 8th Street in downtown Philadelphia finds a number of Philadelphia Rapid Transit Company streetcars and crowded sidewalks in this postcard dated October 16, 1908. In the distance is the magnificent Philadelphia City Hall, which interrupts Market Street between 15th Street and Juniper Street. Market Street continues westward, crossing the Schuylkill River through West Philadelphia.

Two streetcars are passing on Market Street at 15th Street with the west entrance of Philadelphia City Hall in the background. The bridge spanning Market Street led directly from the train floor of Broad Street Station of the Pennsylvania Railroad on the left side to the large office building on the right side of the street, allowing passengers an easy crossing from the congested street. Broad Street Station passenger service began on June 5, 1881 and ended on April 27, 1952. Pennsylvania Railroad's 30th Street Station became its main Philadelphia station.

PHILADELPHIA'S STREETCAR HERITAGE

KENNETH C. SPRINGIRTH

AMERICA THROUGH TIME®
ADDING COLOR TO AMERICAN HISTORY

Peter Witt type streetcar No. 8534 is on a 1979 rail excursion along Southeastern Pennsylvania Transportation Authority (SEPTA) Route 50 on Rising Sun Avenue at Reese Street. This was one of 535 single-end cars (Nos 8000–8534) built by J. G. Brill Company between 1923 and 1926 in three separate orders. Powered by two General Electric type 275A motors and weighing 36,620 pounds, the car was delivered in May 1926. (*Clifford R. Scholes collection*)

On the top portion of cover: Route 23 car No. 2750 is at the Chestnut Hill loop on October 4, 1992. Delivered by St. Louis Car Company to the Philadelphia Transportation Company in March 1947 as part of an order of seventy-five Presidents' Conference Committee (PCC) cars (Nos 2726–2800), the car was overhauled by SEPTA in June 1986, served as the Chestnut Hill Trolley from 1992 to 1996, went to Brookville Equipment Corporation on February 25, 2002, and was delivered back to SEPTA completely rebuilt as PCCII car No. 2320 on September 9, 2003. (*Kenneth C. Springirth photograph*)

On the bottom portion of cover: In 1973, PCC car No. 2168, in a special white upper and roof, yellow lower section, and purple stripe paint scheme, is at the Luzerne Depot waiting for the next assignment. The Philadelphia Transportation Company received this car from St. Louis Car Company in July 1948 as part of an order of sixty cars (Nos 2141–2200). This car was overhauled by SEPTA in January 1986. It served as the Chestnut Hill trolley from 1992 to 1996, and was sold to the Baltimore Streetcar Museum in 2005. (*Clifford R. Scholes collection*)

Back cover: Blackpool boat car No. 603 is turning from Lancaster Avenue onto Girard Avenue for a trip over route 15 in 1977. This open-top excursion trolley (streetcar) was built in 1934 by English Electric for Blackpool Corporation Transport of Blackpool, England, and was on loan as part of the United States Bicentennial. (*Clifford R. Scholes collection*)

America Through Time is an imprint of Fonthill Media LLC
www.through-time.com | office@through-time.com

First published 2019

ISBN 978-1-63499-148-3

Typeset in Utopia Std
Printed and bound in England

Published by Arcadia Publishing by arrangement with Fonthill Media LLC
For all general information, please contact Arcadia Publishing:

Telephone: 843-853-2070
Fax: 843-853-0044
E-mail: sales@arcadiapublishing.com
For customer service and orders:
Toll-Free 1-888-313-2665

www.arcadiapublishing.com

Contents

Acknowledgments

Numerous pictures were purchased from Clifford R. Scholes. Books that were outstanding reference sources were *PCC Cars of North America* by Harold E. Cox; *Philadelphia Car Routes* by Harold E. Cox; *Surface Cars of Philadelphia 1911–1965* by Harold E. Cox; *Utility Cars of Philadelphia 1892–1971* by Harold E. Cox; *The Fairmount Park Trolley, A Unique Philadelphia Experiment* by Harold E. Cox; and *The Trolley Coach In North America* by Mac Sebree and Paul Ward. *Philadelphia Trolley Tracks* by Mike Szilagyi, with its extensive roster information by Rev. Edward J. Casey, was an outstanding internet source. *SEPTA Fiscal Year 2019 Annual Service Plan* by their Service Planning Department in April 2018 was the source for the fiscal year 2017 ridership information. The terms "streetcar" and "trolley car" both refer to rail cars that operate on track and draw power from an overhead wire. Historically, both terms have been used interchangeably. The term light rail and trolley car can be the same; however, light rail generally has more exclusive right of way and wider stop spacing while a trolley car or streetcar involves travelling in more mixed traffic and closer stop spacing.

On October 16, 2004, remanufactured PCCII car No. 2320 is posing for a picture on Baltimore Avenue at 42nd Street on a rail excursion. This was the first of eighteen PCC cars (Nos 2320–2237) completely rebuilt by Brookville Equipment Corporation for the restoration of streetcar service on Route 15. The car was originally No. 2750 delivered in March 1947 to the Philadelphia Transportation Company. It was overhauled by SEPTA in June 1986, served as the Chestnut Hill trolley from 1992 to 1996, and was rebuilt as No. 2320 by Brookville Equipment Corporation with delivery to SEPTA on September 9, 2003. Each car was painted green and cream with a narrow maroon band just below the windows almost identical to the paint scheme used on the PCC cars by the Philadelphia Transportation Company (PTC) plus a modified "wing" logo with the SEPTA "S" in place of "PTC" lettering. Residential parking concerns with 59th Street residents resulted in storing the eighteen completed cars at Callowhill Depot, with occasional use for special charters on the Subway Surface routes. The cars went into Route 15 service on September 4, 2005. (*Kenneth C. Springirth photograph*)

Introduction

Philadelphia's first horse-drawn street railway was opened by the Frankford and Southwark Philadelphia City Passenger Railroad Company on January 20, 1858 from 3rd and Berks to 5th and Morris. Track gauge was 5 feet 2.25 inches. This was followed by the West Philadelphia Passenger Railway Company beginning horse car service from 8th and Market to 41st and Haverford. The Philadelphia and Darby Railroad Company opened a horse line on December 24, 1858 from 40th and Woodland to Cobbs Creek, which by January 22, 1859 operated from Market and Woodland to 9th and Main in Darby. Many other companies opened horse car lines. A cable car line began service from 23rd and Columbia to Franklin and Master on January 10, 1885 by the Philadelphia Traction Company. The first electric streetcar line was opened on December 15, 1892 in Philadelphia by the Philadelphia Traction Company on Catharine and Bainbridge Streets. A number of street railway companies were merged into the Union Traction Company, which was incorporated on September 6, 1895. Horse car service ended in Philadelphia on the Callowhill Street line on January 14, 1897. To achieve a unified system, streetcar lines were merged into the Philadelphia Rapid Transit Company (PRT) on May 1, 1902, with 2,932 streetcars operating on 475 miles of track. At its peak in 1911, PRT operated 3,399 streetcars on 678 miles of track. Following the strike of 1910, PRT faced financial problems and came under the Stotesbury-Mitten management. During 1911 to 1913, there were 1,500 new single-end double truck streetcars (Nos 6000–7499) purchased, which was the largest fleet of standardized streetcars ever purchased by one transit company. Beginning October 15, 1911, each streetcar line was assigned a number, beginning with the 16th Street line which became Route 2. Between 1914 and 1918, five streetcar lines were abandoned: Route 68 (Point Breeze – 13th and 15th Streets); Route 85 (Race and Arch Streets); Route 82 (Chester Avenue – 42nd Street); Route 67 (Zoo line – 33rd and 36th Streets); and Route 84 (Lancaster Avenue – Arch Street).

During 1918 and 1919, there were 240 "Hog Island" cars (Nos 4000–4129 and 5000–5109) purchased. In 1919, twenty-five Birney cars (Nos 1000–1024) were purchased, followed by ten cars purchased second hand in 1920 (renumbered Nos 1025–1034) and five new cars (Nos 1–5) in 1922.

The Philadelphia Rural Transit Company, a PRT subsidiary, began operating twenty buses on a 9.7-mile route on Roosevelt Boulevard in 1923. On October 14, 1923, PRT began operating a 5.6-mile trackless trolley line on Oregon Avenue using nine "Rail-less" vehicles that looked like motorized Birney cars. These were replaced in 1935 with eight thirty-seat trackless trolleys. In 1923, there were 135 double-end cars (Nos 5200–5334) purchased, followed by 535 single-end cars (Nos 8000–8534) known as "eighty hundreds" during 1923 to 1926. After 1926, there were no purchases of streetcars until 1938.

The 1920s and 1930s were difficult times for PRT and for streetcar companies in the United States, because of the Depression. Between 1921 and 1938, nineteen streetcar routes were taken off, as follows: Route 27 (Kensington – 12th and 13th Streets); Route 80 (Point Breeze Avenue); Route 28 (North 2nd and 3rd Streets); Route 45 (Hog Island); Route 44 (Lansdowne Avenue – Arch Street); Route 23B (Chestnut Hill Loop – City Line); Route 23A (Chestnut Hill Loop – Wheel Pump Loop); Route 24 (York Road – 15th and 16th Streets); Route 16 (North 19th and 20th Streets); Route 19 (Luzerne – 6th and 7th Streets); Route 49 (Wayne – Chelten Avenue); Route 51 (Columbia Avenue – 8th and 9th Streets); Route 35 (52nd Street – Parkside), Route 4 (Tioga – 6th and 7th Streets); Route 22 (Doylestown); Route 74 (Hatboro); Route 76 (Darby – Chester); Route 77 (Chester – Media); and Route 71 (Darby – Media).

The first modern Presidents' Conference Committee (PCC) cars for Philadelphia were twenty cars (Nos 2001–2020) ordered by PRT from St. Louis Car Company in March 1938. They were delivered in August 1938. One car was displayed at City Hall in downtown Philadelphia, and another PCC car was placed at Chelten and Pulaski Avenues for Route 53 riders to view. The first PCC cars were placed in Route 53 revenue service at 5:19 a.m. on August 14, 1938. Three Brilliners (Nos 2021–2023) were purchased from the J. G. Brill Company in 1939.

As ridership dropped during the Depression, PRT was not able to earn the rental payments required by the consolidation agreements of 1895–1902. To protect its security owners, preserve the company, and continue serving the public; PRT petitioned the United States District Court for the Eastern District of Pennsylvania on October 1, 1934 to have the court assume jurisdiction over the company. The petition was approved, and a reorganization plan was filed on December 4, 1934. After many court hearings, along with approval of City Council and the Pennsylvania Public Utility Commission, a reorganization plan was adopted, creating the Philadelphia Transportation Company effective January 1, 1940.

Left: A PRT streetcar is on Arch Street (Originally Mulberry Street, it was renamed Arch in 1854) east from 15th Street in this busy 1910 postcard scene in Center City Philadelphia with pedestrians and horse-drawn vehicles. The left side of the street has the tall United Gas Improvement Company (UGI) building that was built in 1899. In 1974, UGI relocated their corporate headquarters to Valley Forge, Pennsylvania. On the right side is the spire of the Greek Revival-style Arch Street Methodist Church completed in 1870.

Below: Two PRT streetcars are passing on Market Street between 11th and 12th Streets in Center City Philadelphia in this postcard scene around 1920. Snellenburg's Department Store occupied an entire block on Market Street between 11th and 12th Streets. Founded in 1869 by Joseph Snellenburg, the store moved to the 12th and Market Streets location in 1889. Snellenburg's opened its first suburban store in Willow Grove in October 1953. Its downtown Philadelphia store closed in 1962, and the suburban stores were sold to Lit Brothers.

Above: Horse-drawn cars are sharing the roadway with a PRT streetcar on Market Street (originally known as High Street) west from 11th Street in Center City Philadelphia in this postcard scene postmarked October 16, 1908. December 28, 1957 was the last day of streetcar service on Market Street, with Route 17 converted to bus operation. Market Street has continued to be one of the principal locations for business in Philadelphia.

Right: Two PRT streetcars are on Chestnut Street in Center City Philadelphia in this postcard postmarked October 14, 1910. Originally named Wynne Street, it was renamed Chestnut Street by William Penn in 1684. Route 13 was the last streetcar line using this portion of Chestnut Street until it was rerouted into the subway on September 9, 1956.

Two PRT streetcars are passing on Market Street in front of the Reading Terminal in this 1910 postcard scene. This Italian Renaissance-style station (characterized by the elaborate roof cornice and arched windows) was opened on January 29, 1893 by the Philadelphia & Reading Railroad, and the last train to use it was on November 6, 1984. This station was replaced by the Market East Station (as part of the Center City Connector), which was renamed Jefferson Station on September 4, 2014. In 2019, the former Reading Terminal building is used for commercial purposes.

Two PRT Market Street Elevated trains (consisting of type M-1 cars built by Pressed Steel Company during 1906 to 1911) are passing at the 40th Street Station (located at the border between Spruce Hill and Powelton Village neighborhoods of West Philadelphia) in this postcard postmarked February 15, 1911. The line opened on March 4, 1907 from 69th Street Terminal and entered the subway at 23rd Street to loop around City Hall at 15th Street in downtown Philadelphia.

1

From PTC to SEPTA

PRT and some smaller transit companies merged into The Philadelphia Transportation Company (PTC) on January 1, 1940. In June 1940, there were 130 PCC cars (Nos 2501–2580 and 2031–2080) purchased. Every PCC car that was purchased new in Philadelphia was built by St. Louis Car Company.

With the exception of Route 61 (Manayunk)—replaced by buses on August 9, 1941 and by trackless trolleys on October 5, 1941—during World War II streetcar routes were not replaced by buses. During 1942, 110 PCC cars (Nos 2081–2090 and 2581–2680) were purchased. In 1945, PTC had 1,898 streetcars, 677 buses, and seventy-four trackless trolleys. Another 100 PCC cars (Nos 2701–2800) were received between February and April 1947 for Route 23. Replacing the slow vintage 6000 series Nearside cars, the first day of service for the 2700 series PCC cars on Route 23 was May 11, 1947. Route 23 daily ridership declined to 100,000 per day and eighty-five of the 2700 series cars were assigned to Route 23. The fifteen remaining cars went to Callowhill Depot for Routes 13 and 43. The final purchase of new PCC cars was in 1948 for 110 (Nos 2091–2200). PTC had 450 PCC cars in service by 1948.

From 1947 to 1953, streetcars were removed from eleven routes as follows: Route 78 (Lansdowne); Route 14 (42nd Street); Route 29 (Tasker and Morris Streets); Route 73 (Frankford and Bridesburg); Route 75 (Wyoming Avenue); Route 38A (52nd Street Extension); Route 25 (Cedar and Memphis Streets); Route 59 (Bustleton); Route 55 (York Road); Route 65 (York Road - 6th and 7th Streets); and Route 18 (South 22nd and 23rd Streets). Between 1921 and 1953, streetcar lines declined by thirty-one.

In 1955, National City Lines (NCL) obtained control of PTC. There was a massive conversion of streetcar lines to bus operation. Between May 29, 1955 and December 29, 1957, streetcar service ended on the following thirty-two routes: Route 7 (Strawberry Mansion - 22nd and 23rd Streets); Route 70 (52nd Street - Bala); Route 54 (Lehigh Avenue); Route 57 (Jefferson and Master Streets); Route 63 (Catharine and Bainbridge Streets); Route 66 (Holmesburg); Route 81 (Passyunk Avenue); Route 38 (Baring Street - Subway); Route 37 (Westinghouse); Route 5 (Frankford - 2nd and 3rd Streets); Route 33 (North 22nd Street); Route 26 (Olney Avenue); Route 9 (Strawberry Mansion); Route 48 (Fairmount - Arch Street); Route 52 (Chelten - Midvale); Route 3 (Frankford - Columbia); Route 8 (Dauphin Street and Susquehanna Avenue); Route 39 (Dauphin Street and Susquehanna Avenue); Route 21 (North 17th and 18th Streets); Route 64 (Federal and Wharton Streets); Route 31 (Market and 63rd Streets); Route 43 (Spring Garden Street); Route 12 (Grays Ferry); Route 40 (Lombard and South Streets); Route 42 (West Spruce Street); Route 79 (Snyder Avenue); Route 2 (15th and 16th Streets); Route 41 (63rd Street); Route 46 (58th and 60th Streets); Route 17 (South 19th and 20th Streets); Route 20 (12th and 13th Streets); and Route 32 (South 17th and 18th Streets).

In the late 1950s, a number of 2700 series PCC cars were placed in storage. Route 6 was converted to bus operation north of City Line Loop (Cheltenham Avenue) to Willow Grove, with streetcars operating from Broad and Olney to City Line Loop on June 8, 1958. In 1966, the stored all-electric 2700 series PCC cars were sent to the Courtland Shop and returned to service at Luzerne Depot replacing the earlier PCC (air) cars that were equipped with air-controlled doors, brakes, and wipers. The southern portion of Route 47 became bus operated from 8th Spring Garden Streets to 8th and Wolf Streets on July 9, 1967.

Southeastern Pennsylvania Transportation Authority (SEPTA) purchased and took over PTC on September 30, 1968. SEPTA converted the remaining northern portion (Spring Garden Street to Godfrey Avenue) of the Route 47 streetcar line to bus operation on June 15, 1969. In 1969, Luzerne Depot's 2100 and 2700 series PCC cars were exchanged for the PCC air cars at Callowhill and Woodland Depots in preparation for the replacing of the Girard Avenue Bridge. That bridge construction severed the only track connection between the West Philadelphia and Luzerne streetcar routes until it reopened in June 1972. SEPTA needed the best streetcars for the Subway Surface Lines. Route 62 was discontinued, and certain Route 13 trips were extended from Yeadon to Darby beginning January 24, 1971.

During 1973, PCC cars Nos 2168 and 2565 were rebuilt, with car No. 2168 painted lemon yellow with a purple belt rail and white top and car No. 2565 painted orange with blue trim and a white top. Additional PCC cars received the orange, blue and white paint scheme.

The loss of sixty streetcars at the Woodland Depot fire of October 23, 1975 created a severe shortage of streetcars in West Philadelphia. All operable 2700 series PCC cars were transferred from Germantown Depot to Woodland Depot on November 6, 1975 for the Subway Surface Lines. SEPTA began a General Overhaul (GOH) Program in 1979 on 112

PCC cars of 1947 and 1948 vintage, beginning with No. 2715 at approximately $100,000 per car.

Buses took over Route 60 (Allegheny – Richmond) on September 4, 1977; Route 50 (Lawndale – 4th and 5th Streets) on October 12, 1980; Route 53 (Wayne – Erie) on May 16, 1985; Route 6 (City Line – Broad and Olney) on January 12, 1986; Route 23 (Chestnut Hill) on February 21, 1992; Route 56 (Erie – Torresdale) on June 14, 1992; and Route 15 (Girard Avenue) on September 13, 1992.

Kawasaki Heavy Industries built 112 single-end streetcars (light rail vehicles) (Nos 9000–9111) for Routes 10, 11, 13, 34, and 36 for the City Division. The cars were designed to handle the subway surface tunnel curves and the 33-foot radius of Philadelphia's surface trackage. Beginning October 10, 1980, car No. 9000 entered service. As a result of testing, modifications were made to the remaining cars which were all in service before the end of 1982.

The new Elmwood Depot at 7311 Elmwood Avenue in Southwest Philadelphia opened on November 20, 1981 to store and maintain the new Kawasaki cars for SEPTA Routes 11, 13, 34, and 36. Costing $23 million, including land acquisition, the Elmwood Depot has the capacity to store 120 streetcars. The new Woodland Shop at the site of the former Woodland Depot was dedicated on June 25, 1984, and handled major repairs of city and suburban Kawasaki cars. Trackage on 58th Street between Chester and Woodland Avenues that was used by Route 46 until it was converted to bus operation on August 11, 1957 was uncovered and reconnected in 1995. This trackage was later used to divert Route 13 streetcars from Chester Avenue to Woodland Avenue because of bridge reconstruction.

Beginning September 13, 1992 and ending on June 15, 1996 because of budget cuts, SEPTA operated "The Chestnut Hill Trolley" using two PCC streetcars, providing weekend service every twenty minutes on the 2-mile northernmost portion of Route 23 between the Chestnut Hill Loop and Germantown Depot.

A free "Holiday Trolley" on vintage trolleys was operated by SEPTA from November 29 to December 29, 1996 every twenty minutes on weekdays from 10:30 a.m. to 7:30 p.m. and on weekends from 10 a.m. to 5 p.m. on 11th and 12th Streets from Noble to Bainbridge Streets. The free trolley was sponsored by the City of Philadelphia, the Marriott Hotel, the Pennsylvania Convention Center, and SEPTA. In a later year, SEPTA operated a Holiday Trolley for a 25-cent fare, using vintage trolleys between Arch and Bainbridge Streets and modern trolleys on a Market to Market route from Arch Street to Snyder Avenue. A "Historic Holiday Trolley" was operated from November 24, 2000 to December 31, 2000 on Thursdays through Sundays between 10 a.m. and 5 p.m. on former streetcar Route 23 trackage on 11th and 12th Streets in Center City Philadelphia.

Service was about every twenty minutes. Certain trips operated into the neighborhoods of South Philadelphia. The fare was either a valid SEPTA pass or a cash fare of 50 cents.

SEPTA agreed to restore streetcar service on Route 15. Although plans called for new cars, financial concerns resulted in a November 2001 decision to have Brookville Equipment Corporation of Brookville, Pennsylvania, rebuild eighteen PCC cars for $23 million, which included rebuilt bodies, new trucks, addition of wheel chair lift at the center door, modern control system, and installing air conditioning. Streetcar service was restored on Route 15 on September 4, 2005 using rebuilt PCC cars known as PCC-II cars. From September 13, 1992 to September 3, 2005 when buses operated on Route 15, Route 10 streetcars were stored and maintained at the Elmwood Depot.

In 2019, SEPTA's Kawasaki streetcars have been in service about seven years longer than their thirty-year design life, and plans are being formulated for the next generation of streetcars. The Delaware Valley Regional Planning Commission released a report, "Modern Trolley Station Design Guide," dated December 2017, on SEPTA City Transit Division Routes 10, 11, 13, 15, 34, and 36. The report noted that modernizing the trolley system will be challenging, expensive, and include extensive coordination between SEPTA, the City of Philadelphia, stakeholder agencies, and communities along trolley corridors. In 2018, five of Philadelphia's city routes (10, 11, 13, 34, and 36) are served by Kawasaki cars, and are known as "subway surface" routes because they run in a tunnel in center city, and emerge from the tunnel operating on the street as a streetcar. Route 15 (served by PCC II streetcars) runs entirely on the street as a streetcar. SEPTA is planning to purchase new low-floor vehicles with multi-door boarding.

Where necessary, SEPTA has renewed streetcar trackage. From March 18, 2018 to June 9, 2018, a portion of Route 36 trackage was replaced on Elmwood Avenue between 59th and 65th Streets. The tracks, installed in the 1970s, had deteriorated because of weather, traffic, and the road surface age. During the construction period, buses replaced Route 36 streetcars from 80th Street and Eastwick Avenue to the 40th Street Subway Portal where passengers boarded any Route 11, 13, and 34 streetcar to reach Center City.

For the fifth consecutive summer, from 10 p.m. on Friday July 13, 2018 through 4 a.m. on Monday July 23, 2018, SEPTA closed the streetcar tunnel for a nine-day, around-the-clock power, track, and station improvement projects. SEPTA's Track Department replaced rail where necessary and cleared clogged track drains throughout the tunnel. Streetcar Routes 10, 11, 13, 34, and 36 used their on the street diversion trackage to connect with SEPTA's Market – Frankford Line at 40th and Market Streets. Passengers boarded the Market – Frankford Line at the 40th Street Station for travel to and from Center City Philadelphia.

Above: Philadelphia Transportation Company (PTC) car No. 2510 is westbound on Route 42 on Walnut at 6th Streets in 1956. Delivered in December 1940, this was one of eighty PCC cars (Nos 2501–2580) (built by St. Louis Car Company) powered by four General Electric type 1198F3 motors, seated fifty passengers, and weighed 36,000 pounds. Route 42 was converted to bus operation on September 9, 1956. The car was scrapped in 1982. (*Clifford R. Scholes collection*)

Right: In 1956, double-end car No. 5333 is at the 60th Street and Lansdowne Avenue northern terminus of Route 46 in the Carroll Park neighborhood of West Philadelphia. Delivered in August 1923 as car No. 5317 and later renumbered 5333, the car was modernized in October 1941. This was one of 135 cars (Nos 5200–5334) built by J. G. Brill Company. Powered by two General Electric type 275 motors, the car weighed 37,720 pounds. Buses took over Route 46 on August 10, 1957, and the car was scrapped that same month. (*Kenneth C. Springirth photograph*)

Route 2 car No. 8373 is at 15th and Pollock Streets in the Girard Estates neighborhood of South Philadelphia waiting for departure time in 1956. Delivered in December 1923 as a two-man car, the car was rebuilt as one-man/two-man car in 1928. Powered by two General Electric type 275A motors and weighing 35,820 pounds, this was one of 385 cars (Nos 8000–8384) built by J. G. Brill Company. Route 2 was converted to bus operation on November 4, 1956. Streetcar No. 8373 was scrapped in August 1957. (*Clifford R. Scholes collection*)

Market Street on trackage around City Hall in Center City Philadelphia is the location of Route 32 car No. 8300 in 1956. This car was delivered in November 1923 and was scrapped in February 1958. Route 32 streetcar service ended December 28, 1957 with partial replacement by bus Routes 2 and 33. (*Clifford R. Scholes collection*)

In 1955, PTC car No. 8165 (delivered in October 1923 and scrapped in August 1957) is in the lineup of Route 20 cars near the Municipal Stadium in South Philadelphia on the east side of the southern end of Broad Street, probably for an Army-Navy football game. Route 20 was merged into Route 23 on December 29, 1957, and the trackage south of Bigler Street was abandoned. Opened on April 15, 1926, the stadium seated 102,000 people. It was demolished on September 23, 1992 because of structural and fire hazard concerns. (*Clifford R. Scholes collection*)

Route 20 car No. 8514 is northbound turning from Broad Street onto Terminal Avenue near the Municipal Stadium in 1956. This streetcar was delivered in May 1926, remodeled in September 1940, and was scrapped in March 1958. (*Clifford R. Scholes collection*)

Luzerne Depot, on April 27, 1968, is the location of PCC car No. 2535 (delivered in January 1941 and scrapped in March 1982) and No. 2562 (delivered in January 1941 and scrapped in 1982). Opened in 1912, Luzerne Depot housed 335 streetcars inside and 115 streetcars in the outside yard. In 1968, it was home to Routes 6, 47, 50, 53, 56, and 60. Route 23 streetcars were moved to Luzerne Depot in September 1976. During 1992, Luzerne Depot housed only buses, and completely closed in April 1997. (*Kenneth C. Springirth photograph*)

PCC car No. 2168 is turning into Mount Moriah Loop on Kingsessing and Cemetery Avenues on a June 8, 2003 rail excursion, with a special destination sign showing a different location. This loop is the location of Mount Moriah Cemetery, and in 2018, certain rush-hour trips terminate here. The cemetery was incorporated by an act of the Pennsylvania State Legislature in 1855. After years of neglect, the cemetery closed in 2011. The Friends of Mount Moriah Cemetery Inc., a nonprofit group, organized to preserve and make the cemetery as accessible as possible because of its history, ecology, and important community resource. (*Kenneth C. Springirth photograph*)

The Woodland Depot after the October 23, 1975 fire looked like a war zone. The loss of sixty streetcars in that fire caused havoc on SEPTA's streetcar operation. Although thirty PCC cars were purchased from the Toronto Transit Commission, some of those cars had major issues. This may have been a factor in the conversion of Route 60 to bus operation on September 4, 1977. (*Clifford R. Scholes collection*)

The No. 603 "Boat Trolley" from Blackpool, England, is on a low-bed truck trailer in Philadelphia for delivery to SEPTA to celebrate the United States Bicentennial. It was placed in service on a portion of Route 50 on August 2, 1976. (*Clifford R. Scholes collection*)

In 1978, PCC car No. 2318 is heading a line of cars at Luzerne Depot. Powered by four Westinghouse type 1432 motors, this was originally Birmingham Transit Company car No. 846 one of forty-eight cars (Nos 800–847) built by Pullman-Standard Car Manufacturing Company and delivered between November 1946 and January 1947. In 1952, it was one of forty-eight cars acquired by the Toronto Transportation Commission (TTC), becoming car No. 4746 of the series (Nos 4700–4747). It was acquired by SEPTA in March 1976, renumbered 2318, and scrapped in September 1981. (*Clifford R. Scholes collection*)

Training trolley No. 2780 is at Luzerne Depot in 1979. Delivered to PTC in April 1947, the car was overhauled by SEPTA in August 1981 and was sold to Brookville Equipment Corporation for parts in 2004. (*Clifford R. Scholes collection*)

On February 3, 2007, this unique streetcar style subway entrance at 37th and Spruce Streets is in the center of the University of Pennsylvania campus. Funded by the University of Pennsylvania Class of 1956, the Gomaco Trolley Company began fabricating the entrance in July 2006, and it was delivered in mid-October 2006. It was unveiled on October 28, 2006 at 10 a.m. (*Kenneth C. Springirth photograph*)

The streetcar-style subway entrance stairway for eastbound streetcar (trolley) Routes 11, 13, 34, and 36 to Center City Philadelphia is awaiting passengers in this February 3, 2007 view. PRT and later PTC operated streetcars on Woodland Avenue through the University of Pennsylvania campus. Effective September 11, 1955, PCC cars were assigned to all of the subway-surface lines. On November 5, 1955 at 5:07 a.m., Route 36 PCC car No. 2628 was the first westbound streetcar through the new extended subway to 40th Street. (*Kenneth C. Springirth photograph*)

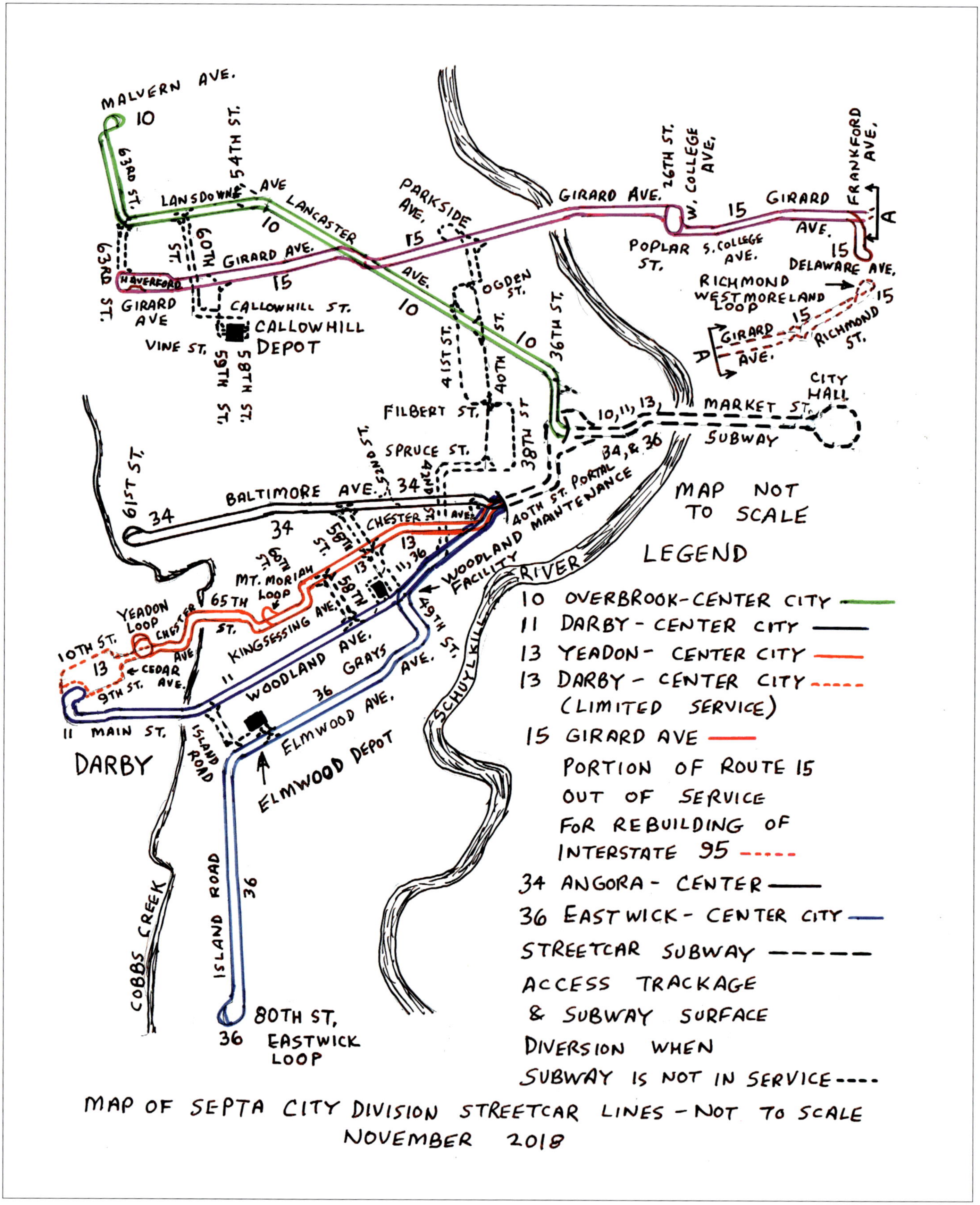

This map shows SEPTA's six streetcar lines in the City of Philadelphia. The five subway surface lines are Routes 10, 11, 13, 34, and 36. Route 11 crosses into Darby. Route 13 serves Yeadon, and certain trips are extended into Darby. Route 15 operates entirely on the street.

2

Routes 6 and 10

Route 6 (Willow Grove) was opened September 15, 1924 between City Line and Ogontz Avenue to Willow Grove via Limekiln Pike by the Philadelphia Rapid Transit Company (PRT). On March 3, 1929, the line was extended south via Ogontz Avenue to Broad and Olney Avenue. From Broad and Olney (transfer point for Broad Street Subway), Route 6 primarily operated on Ogontz Avenue in Philadelphia. Crossing the City Line at Cheltenham Avenue, the line used the private right of way median of Limekiln Pike, proceeded onto Keswick Avenue in Glenside, and was on the side of the road private right of way to Willow Grove Park. With portions of the Limekiln Pike median acquired by the Pennsylvania Highway Department, the Philadelphia Transportation Company (PTC) replaced streetcars with buses between Willow Grove and City Line Avenue effective June 8, 1958 and operated streetcars from City Line Avenue to Broad and Olney Avenue. The Southeastern Pennsylvania Transportation Authority replaced the remainder of Route 6 with buses on January 12, 1986.

Route 10 (Overbrook - Center City) was established by PRT on December 15, 1906 from Belmont and Parkside to the 15th Street Subway Station, and the Juniper Street Subway Station opened on August 3, 1908. On June 20, 1926, the line was extended to operate from 63rd and Malvern to the Juniper Street Subway Station. From the subway surface tunnel under Market Street, Route 10 exits the subway at the 36th Street Portal, goes north on 36th Street, turns northwest onto Lancaster Avenue, intersects Girard Avenue (where the Route 15 streetcar briefly intersects before turning left to continue on Girard Avenue), continues on Lancaster Avenue to 52nd Street, turns west onto Lansdowne Avenue, and continues west on Lansdowne Avenue to 63rd Street, where it turns north to the 63rd and Malvern Loop.

Route 6 PCC car No. 2122 (delivered to PTC in September 1948 and overhauled by SEPTA in February 1981) is on Ogontz Avenue at Wyncote Avenue in the Ogontz neighborhood of Upper Northern Philadelphia on May 31, 1968. One of fourteen PCC cars sold to the San Francisco Municipal Railway (SFMR) for F-line service, the car was restored by Morrison-Knudsen Corporation in 1993 and renumbered 1055, with a green, cream, and maroon stripe paint scheme. (*Kenneth C. Springirth photograph*)

On May 31, 1968, northbound Route 6 PCC car No. 2519 (delivered in December 1940 and scrapped in 1973) is turning from Old York Road to Champlost Street in the Fern Rock neighborhood of Upper Northern Philadelphia. (*Kenneth C. Springirth photograph*)

Broad Street at Grange Avenue is the location of Route 6 PCC car No. 2798 on May 31, 1968. This car was delivered to PTC in April 1947. It was overhauled by SEPTA in September 1985 and was sent to Brookville Equipment Corporation on July 15, 2002 for rebuilding into PCCII car No. 2323 that was delivered to SEPTA on March 11, 2004. (*Kenneth C. Springirth photograph*)

Southbound Route 6 all-electric PCC car No. 2122 is on tree-lined Park Avenue at Grange Avenue passing classic Philadelphia row homes on May 31, 1968. The all-electric PCC cars (electrically operated brakes, doors, and windshield wipers) were cars Nos 2091–2290 and 2701–2800. (*Kenneth C. Springirth photograph*)

On May 31, 1968, Route 6 PCC car No. 2126 is turning from Old York Road to Nedro Avenue for a southbound trip to the Broad and Olney terminus. This car was delivered to PTC in September 1948 and was overhauled by SEPTA in January 1982. The front end was damaged in June 1990. It was sold to the San Francisco Municipal Railway for parts in October 1992. (*Kenneth C. Springirth photograph*)

In an attractive Bicentennial paint scheme, SEPTA Route 6 PCC car No. 2288 (scrapped in 1981) is on Ogontz Avenue at 78th Street in 1976. This car was originally car No. 791, one of seventy-five cars (Nos 725–799) built by St. Louis Car Company for Kansas City Public Service Company and delivered between March and August 1946. There were forty of these cars sold to PTC and delivered between December 1954 and July 1955. They were refurbished and renumbered 2251–2290. (*Clifford R. Scholes collection*)

On July 27, 1979, SEPTA car No. 2054 is at the Route 6 Broad Street and Olney Avenue southern terminus. Delivered in February 1941, seating fifty passengers, and powered by four Westinghouse type 1432D motors, this was one of ten PCC cars (Nos 2031–2040) ordered in June 1940 by PTC from St. Louis Car Company. This car, slightly different than the first PCC cars, was painted silver with a blue stripe to commemorate the fortieth anniversary of the first PCC cars (Nos 2001–2020) delivered to PRT in July 1938. (*Kenneth C. Springirth photograph*)

The City Line Loop northern terminus of Route 6 is the location of PCC car No. 2054 on July 27, 1979. This car was acquired by the Electric City Trolley Museum in Scranton, Pennsylvania, in 2005. (*Kenneth C. Springirth photograph*)

In 1979, Route 6 PCC car No. 2245 is on Park Avenue at Grange Avenue. This car was originally Kansas City Public Service Company car No. 765, built by St. Louis Car Company in 1946. It was purchased by the Toronto Transportation Commission in 1957 and renumbered 4762. SEPTA acquired the car in March 1976, renumbered it 2245, and it was scrapped in February 1982. (*Clifford R. Scholes collection*)

In 1979, Park Avenue at Chew Avenue is the location of Route 6 PCC car No. 2283 in a Bicentennial paint scheme honoring the State of New York. This car was originally Kansas City Public Service Company car No. 734 built by St. Louis Car Company in 1946. It was acquired by PTC in June 1955, refurbished plus renumbered 2283 in August 1955, and sold in November 1985 to the Vintage Electric Streetcar Company of Windber, Pennsylvania. (*Clifford R. Scholes collection*)

Still in its Toronto paint scheme, Route 6 PCC car No. 2304 is southbound on Park Avenue at Grange Avenue in 1980. This was originally Birmingham Transit Company car No. 810 built by Pullman-Standard Car Manufacturing Company and delivered with forty-seven other cars between November 1946 and January 1947. It was acquired by the Toronto Transportation Commission in 1952 where it was renumbered 4710. SEPTA purchased it in April 1976, renumbered it 2304, and it was scrapped in March 1982. (*Clifford R. Scholes collection*)

In August 1983, Route 6 PCC car No. 2124 is at the City Line Loop waiting for departure time. This car was delivered to PTC in August 1948, was overhauled by SEPTA in August 1980, and was sold to San Francisco Municipal Railway where it was restored by the Morrison Knudson Corporation and renumbered 1058 in 1993. Honoring Chicago's streetcars, the car was repainted in 2010 to the Mercury Green, Croydon Cream, and Swampholly Orange that was used on Chicago streetcars after World War II. (*Kenneth C. Springirth photograph*)

Route 10 PCC car No. 2168 is posed for a picture stop on route 10 at the 36th and Ludlow Streets Subway Portal on an Electric Railroaders' Association Convention rail excursion on July 7, 1973. (*Kenneth C. Springirth photograph*)

On February 3, 2007, Route 10 Kawasaki car No. 9103 (delivered on March 22, 1982 and placed in revenue service on March 26, 1982) is at the 63rd Street and Malvern Avenue Loop in the Overbrook neighborhood of the northwestern part of West Philadelphia. Kawasaki Heavy Industries built 112 single-end streetcars (Nos 9000–9111) for subway surface lines 10, 11, 13, 34, and 36. All of these cars were received before the end of 1982. (*Kenneth C. Springirth photograph*)

Route 10 Kawasaki car No. 9077 is turning from 36th Street to shortly enter the streetcar tunnel for an eastbound trip through the subway to the 13th and Juniper Streets Subway Station. The 50-foot-long Kawasaki cars featured an automatically controlled comfort system for air conditioning and heating. Each car weighed 57,882 pounds, seated fifty-one passengers, and had a maximum speed on level track of 50 mph. In fiscal year 2017, the 5.9-mile Route 10 had an average weekday ridership of 11,163 and required seventeen cars during rush hours. (*Kenneth C. Springirth photograph*)

3

Routes 11 and 13

Route 11 (Darby - Center City) was opened as a subway surface line on December 15, 1906 from 9th and Main in Darby to the 15th Street Subway Station. On August 3, 1908, the Juniper Street Subway station was open to passengers. From the subway surface tunnel under Market Street, Route 11 exits the subway at the 40th Street Portal and operates on Woodland Avenue crossing Darby Creek entering Darby on Main Street. The line crosses the CSX Transportation Philadelphia Subdivision freight line at grade and continues west on Main Street with a turn north on 9th Street to enter the Darby Transportation Center.

Route 13 (Yeadon - Center City with some trips extended to 9th and Main in Darby) operated from Front and Chestnut to 65th and Kingsessing on January 24, 1909. Beginning May 25, 1913, it provided a through trip from Front and Chestnut to 9th and Main. On September 9, 1956, the line operated from the Juniper Street Subway Station to Yeadon Loop. Effective January 24, 1971, Route 62 was discontinued, and certain Route 13 trips were extended to Darby. From the subway surface tunnel under Market Street, Route 13 exits the subway at the 40th Street Subway Portal and runs southwest on Woodland Avenue a short distance turning onto Chester Avenue. The line turns left on 60th Street, turns right to head southwest on Kingsessing Avenue, turns north onto 65th Street, and turns on to another part of Chester Avenue to Yeadon Loop. Some streetcars operate west of Yeadon Loop on Chester Avenue to Cedar Avenue. Westbound cars turn right onto Cedar Avenue, left on 10th Street, and left on Main Street to enter the Darby Transportation Center. Eastbound cars leave the Darby Transportation Center making a left turn onto 9th Street, a brief jog on Summit Street, back on 9th Street, and a brief jog on Cedar Street to reach Chester Avenue.

Eastbound Route 11 PCC car No. 2165 (delivered in July 1948 and scrapped in 1985) is on Main Street about to cross the Baltimore & Ohio Railroad at grade in the middle of 6th Street in the Borough of Darby in Delaware County, Pennsylvania, on May 20, 1975. (*Kenneth C. Springirth photograph*)

On May 30, 1970, SEPTA Route 11 car No. 2188 has emerged from the 40th Street Subway Portal for a westbound trip to Darby. This car was delivered in July 1948 and was destroyed in the Woodland Depot fire of October 23, 1975. Route 11 service levels have been fairly stable except for Sunday, where the frequency on August 26, 2018 was every twenty minutes compared to a twelve- to fourteen-minute frequency on January 21, 1968. (*Kenneth C. Springirth photograph*)

Route 11 SEPTA cars Nos 2166, 2160, and a third car are at the Darby Loop on July 27, 1979. Car 2166 was delivered to PTC in July 1948, overhauled by SEPTA in January 1984, and sold to the New Orleans Regional Transit Authority in January 1995 where the body was scrapped. Car 2160 was delivered to PTC in June 1948, overhauled by SEPTA in January 1986, went to the Baltimore Streetcar Museum in 2005, later to Lancaster, Pennsylvania, and in 2012 to the Warren Rail Car in Warren, Pennsylvania. (*Kenneth C. Springirth photograph*)

Route 11 Kawasaki car No. 9003 (delivered June 4, 1981 and placed in revenue service on August 12, 1981) is on Main Street in Darby on July 24, 1999 crossing the CSX Philadelphia Subdivision track, which runs 89 miles between Philadelphia and Baltimore. It dates back to 1886, running somewhat parallel to the Pennsylvania Railroad mainline, which is now the Northeast Corridor used by Amtrak and SEPTA. It gave the former Baltimore & Ohio Railroad access to Philadelphia, and trackage rights with the Reading Railroad gave it access to Jersey City and New York area market. On April 30, 1987, the B&O was absorbed into the Chesapeake & Ohio Railway, which merged into the CSX Transportation on August 31, 1987. (*Kenneth C. Springirth photograph*)

On October 2, 2010, Route 11 Kawasaki car No. 9050 (delivered September 21, 1981 and placed in revenue service on October 2, 1981) in regular service and PCCII car No. 2324 (originally PCC car No. 2777 that was rebuilt by Brookville Equipment Corporation and delivered to SEPTA on March 26, 2004) on a special OC Trolley Fest excursion are at the Darby Terminal. (*Kenneth C. Springirth photograph*)

Passenger tents have been set up on 40th Street at Market Street during the special "Maintenance Blitz" that closed the streetcar subway over a number of days to enable major repairs to be completed. Streetcar Routes 10, 11, 13, 34, and 36 used street diversion trackage to connect with SEPTA's Market - Frankford Line at 40th and Market Streets. Passengers boarded the Market - Frankford Line at the 40th Street Subway Station for travel to and from Center City Philadelphia. On August 15, 2014, Route 11 Kawasaki car No. 9087 (delivered October 28, 1981 and placed in revenue service on November 25, 1981) is on 40th Street crossing Market Street on the diversion route. (*Kenneth C. Springirth photograph*)

Route 11 Kawasaki car No. 9001 (delivered May 26, 1981 and placed in revenue service on July 13, 1981) is on 40th Street turning on to Spruce Street on August 15, 2014 on the diversion route, because the streetcar subway was closed to handle major repairs. By 2018, there have been five "Maintenance Blitzes" held during the summer when there is less ridership. (*Kenneth C. Springirth photograph*)

On August 15, 2014, Route 11 Kawasaki car No. 9095 (delivered on March 10, 1982 and placed in revenue service on March 15, 1982) is on Spruce Street at 40th Street in the University City District of West Philadelphia on the diversion trackage to 40th and Market Streets for passengers to transfer to the Market - Frankford Line. (*Kenneth C. Springirth photograph*)

Route 11 Kawasaki car No. 9066 (delivered on October 14, 1981 and placed in revenue service on October 19, 1981) is heading south on 42nd Street ready to cross Baltimore Avenue on the diversion trackage for a trip to Darby on August 15, 2014. In fiscal year 2017, the 6.7-mile Route 11 had an average weekday ridership of 12,952 and required sixteen cars during rush hours. (*Kenneth C. Springirth photograph*)

Emerging from the 40th Street Subway Portal, PTC Route 13 PCC car No. 2621 in 1964 is operating a trip to Mount Moriah via Chester Avenue, a cutback loop on the Yeadon line. This was one of forty-three cars (Nos 2581–2623) delivered by St. Louis Car Company between May and June 1942. Powered by four General Electric type 1198 motors, each car weighed 35,860 pounds and seated forty-nine passengers. Car No. 2621 was destroyed in the Woodland Depot fire of October 23, 1975. (*Clifford R. Scholes collection*)

On a sunny May 30, 1970, SEPTA Route 13 PCC car No. 2156 is turning from 60th Street to Kingsessing Avenue in the Kingsessing neighborhood of Southwest Philadelphia. This car was delivered to PTC in June 1948, was overhauled by SEPTA in September 1983, and was sold to Brookville Equipment Corporation in 2004. (*Kenneth C. Springirth photograph*)

Amid the blossoming trees, Route 13 Kawasaki car No. 9083 (delivered October 23, 1981 and placed in revenue service on October 27, 1981) is on 10th Street near Union Avenue in Darby, Pennsylvania, in 1982. (*Clifford R. Scholes collection*)

On Chester Avenue at 41st Street in the Woodland Terrace neighbourhood of West Philadelphia, from left to right, three Route 13 Kawasaki cars Nos. 9070 (delivered October 19, 1981 and placed in revenue service on October 22, 1981), 9088 (delivered October 30, 1981 and placed in revenue service on November 9, 1981, and 9032 (delivered April 16, 1981 and placed in revenue service on April 21, 1981) are in view in this 1983 scene. (*Clifford R. Scholes collection*)

On March 28, 1987, Route 13 Kawasaki car No. 9039 (delivered April 30, 1982 and placed in revenue service on May 5, 1982) is at 40th at Market Streets in the Powelton neighborhood of West Philadelphia on the diversion routing when the streetcar tunnel is closed. (*Kenneth C. Springirth photograph*)

Westbound for Darby via Chester Avenue on April 29, 1987, Route 13 Kawasaki car No. 9021 (delivered on May 4, 1982 and placed in revenue service on May 6, 1982) is on Spruce Street at 41st Street. (*Kenneth C. Springirth photograph*)

On June 13, 1992, Route 13 Kawasaki car No. 9043 (delivered September 14, 1981 and placed in revenue service on September 22, 1981) is on Chester Avenue turning into the Yeadon Loop at Callahan Avenue in the Borough of Yeadon. With the large blue area on the side of the cars shown on page 36 difficult to maintain, a narrow stripe below the car windows was used on the revised paint scheme on the car of this page. (*Kenneth C. Springirth photograph*)

Chester Avenue near Cedar Avenue in Yeadon is the location of PCC car No. 2168 on July 24, 1999. This trackage was used by Route 62 until January 24, 1971, when it was replaced by the extension of certain Route 13 trips from Yeadon to Darby. (*Kenneth C. Springirth photograph*)

On February 5, 2006, Route 13 Kawasaki car No. 9042 (delivered October 19, 1982 and placed in revenue service on November 15, 1982) is on Spruce Street at 40th Street in the University City District of Philadelphia. Another revised paint scheme finds the narrow stripe below the car windows has been eliminated when compared with the car on page 37. The car is on diversion trackage to 40th and Market Streets. (*Kenneth C. Springirth photograph*)

Route 13 Kawasaki car No. 9022 (delivered October 22, 1982 and placed in revenue service on December 9, 1982) is leaving the Yeadon Loop for a trip to City Hall in Center City Philadelphia on October 14, 2006. (*Kenneth C. Springirth photograph*)

At Yeadon Loop on June 7, 2009, the cars from left to right are Route 13 Kawasaki car No. 9000 (the first car delivered on July 23, 1980 and placed in revenue service on November 3, 1980); PCC car No. 2194 (delivered to PTC in July 1948, overhauled by SEPTA in April 1988, and later became an overhead line car); and PCCII car No. 2334 (originally PCC car No. 2753 and completely rebuilt by Brookville Equipment Corporation and delivered to SEPTA on September 24, 2004). (*Kenneth C. Springirth photograph*)

On June 7, 2009, PCCII car No. 2334 is on a rail excursion posing for a picture stop on Kingsessing Avenue with the Mount Moriah Cemetery on the other side of the street. This large historic cemetery, located in Philadelphia and across Cobbs Creek in Yeadon, is being preserved by the Friends of Mount Moriah Cemetery, Inc. (*Kenneth C. Springirth photograph*)

Route 13 Kawasaki car No. 9103 is southbound on 40th Street crossing Ludlow Street on the diversion trackage on August 15, 2014. Route 13 service levels have been fairly stable except only weekday rush-hour trips use Mount Moriah Loop, and on Sunday, the frequency on August 26, 2018 was every twenty minutes compared to the twelve-minute frequency on June 21, 1964. (*Kenneth C. Springirth photograph*)

On June 11, 2018, Kawasaki car No. 9013 (delivered October 26, 1981 and placed in revenue service on November 3, 1981) has emerged from the 40th Street Subway Portal for a Route 13 trip to Yeadon via Chester Avenue. In fiscal year 2017, the 7-mile Route 13 had a daily weekday ridership of 12,869 and required eighteen cars during rush hours. (*Kenneth C. Springirth photograph*)

4

Routes 15 and 23

Route 15 (Girard Avenue) began operation on 1895 between Richmond and Norris and 54th and Girard. On April 1, 1902, the line was extended west to 63rd and Girard. It was extended east to Richmond and Allegheny on July 22, 1903. Another extension was made to Bridge and Minor on May 30, 1920. Effective February 24, 1956, the line operated from Richmond and Westmoreland to 63rd and Girard. Buses replaced streetcars on September 13, 1992, with streetcars back in service on September 4, 2005. The rebuilding of Interstate 95 has resulted in Route 15 streetcars temporarily terminating at Frankford and Delaware Avenues since April 29, 2012. The line operates from 63rd Street via Girard Avenue east to where it briefly overlaps with Route 10 on Lancaster Avenue and continues east on Girard Avenue. It goes around the south side of Girard College and rejoins Girard Avenue to Frankford Avenue, where it temporarily heads south to the loop at Frankford and Delaware Avenues. When the Interstate 95 expressway project is completed, Route 15 would continue east on Girard Avenue to Richmond Street to its eastern terminus at Westmoreland Street.

Route 23 (Germantown Avenue – 11th and 12th Streets) operated from 4th and Walnut to Germantown and Hillcrest on July 3, 1896. Beginning August 31, 1913, the line operated from 10th and Ritner to Wheel Pump Loop. On October 31, 1920, the line operated between 10th and Ritner and the Germantown – Bethlehem Loop with an extension south to 11th and Patterson on June 1, 1926. It was merged with Route 20 on December 29, 1957 operating northbound from Bigler Street, north on 10th Street, west on Oregon Avenue, north on 11th Street, east on Huntingdon Street, and north on Germantown Avenue to the Germantown – Bethlehem (Chestnut Hill) Loop.

Southbound Route 23 operated south on Germantown Avenue, south on 10th Street, west on Susquehanna Avenue, south on 12th Street, east on Oregon Avenue, and south on Marvine Street to Bigler Street. Route 23 buses replaced streetcars on February 21, 1992.

On February 28, 1969, SEPTA Route 15 PCC car No. 2130 is on an operator layover at the Richmond and Westmoreland Loop. Car No. 2130 was delivered to PTC in August 1948 and was destroyed in the Woodland Depot fire of October 23, 1975. (*Kenneth C. Springirth photograph*)

Boat trolley No. 603, on loan for the U.S. Bicentennial from Blackpool, England, is on Girard Avenue at 63rd Street, the western terminus of Route 15 in 1977. (*Clifford R. Scholes collection*)

On July 22, 1979, Route 15 PCC car No. 2091 and Peter Witt-type car No. 8534 are at the Richmond and Westmoreland Loop. Car No. 2091 was delivered to PTC in August 1948, overhauled by SEPTA in January 1983, and sold to Gomaco Trolley Company in October 1994. Car No. 8534 was delivered to PTC in May 1926, remodeled in October 1940, taken out of service when streetcar Route 17 was converted to bus operation on December 29, 1957 and Route 32 was eliminated, used for charter service for a number of years, and is now at the Electric City Trolley Museum at Scranton, Pennsylvania. (*Kenneth C. Springirth photograph*)

Route 15 PCC car No. 2114 is on Girard Avenue having just crossed under the Interstate 95 bridge on June 28, 1986. Car No. 2114 was delivered to PTC in August 1948, overhauled by SEPTA in May 1982 and sold to Colorado Springs in June 1995. (*Kenneth C. Springirth photograph*)

On August 31, 1990, Route 15 PCC car No. 2091 is eastbound on Girard Avenue at Front Street, which is a transfer point for the Market - Frankford Elevated Line. This car shows off its classy red, white, and blue paint scheme that SEPTA used on overhauled PCC cars. (*Kenneth C. Springirth photograph*)

The Route 15 western terminus on Girard Avenue at 63rd Street in the Haddington neighborhood of West Philadelphia on September 2, 1990 is the location of PCC car 2713. This car was delivered to PTC in February 1947, overhauled by SEPTA in July 1982, and sold to Brookville Equipment Corporation in 2004. (*Kenneth C. Springirth photograph*)

On May 18, 1992, Route 15 PCC car No. 2711 is at the Richmond and Westmoreland Loop in the Port Richmond neighborhood of Philadelphia. This car was delivered to PTC in February 1947, overhauled by SEPTA in March 1982, and acquired in 2000 by the Pennsylvania Trolley Museum in Washington, Pennsylvania. (*Kenneth C. Springirth photograph*)

The Route 15 eastern terminus at the Richmond and Westmoreland Streets Loop is the location of PCC car No. 2723 on July 13, 1992. Delivered by St. Louis Car Company to PTC in March 1947, this car was overhauled by SEPTA in July 1982 and was acquired by the Pennsylvania Trolley Museum in 1995. (*Kenneth C. Springirth photograph*)

Girard Avenue at Columbia Avenue in the Fishtown neighborhood northeast of Center City Philadelphia is the location of Route 15 PCC car No. 2129 on July 13, 1992. Received by PTC in August 1948, this car was overhauled by SEPTA in November 1982 and was sold to Colorado Springs in May 1995. (*Kenneth C. Springirth photograph*)

Eastbound Route 15 PCC car No. 2710 is on Girard Avenue at Fletcher Street on July 13, 1992. Car 2710 was delivered to PTC in February 1947, overhauled by SEPTA in June 1983, and sold to Gomaco Trolley Company in October 1994. In Philadelphia, PCC cars were used exclusively in regular service from December 29, 1957 (Routes 17 and 32 were the last lines to use the 8000 series cars until December 28, 1957) until October 10, 1980, when the first Kawasaki car No. 9000 arrived. (*Kenneth C. Springirth photograph*)

On July 13, 1992, Route 15 PCC car No. 2097 is on Girard Avenue at Frankford Avenue with the Girard Avenue Station of the Market - Frankford Elevated Line in the background. PTC took delivery of this car in August 1948. Overhauled by SEPTA in July 1981, it was sold to Colorado Springs in June 1995. Route 15 was the last Philadelphia line to use PCC cars until it was converted to bus operation on September 13, 1992. (*Kenneth C. Springirth photograph*)

Route 15 PCC car No. 2095 is turning onto Richmond Street from the Cumberland Street Loop on August 21, 1992 with the elevated Interstate 95 Highway on the left side. St. Louis Car Company delivered this car to PTC in August 1948. It was overhauled by SEPTA in January 1983 and was sold to the Middletown & Hummelstown Railroad in May 1994. (*Kenneth C. Springirth photograph*)

On July 14, 1995, Route 15 Kawasaki car No. 9020 (delivered March 10, 1982 and placed in revenue service on April 15, 1982) is on Girard Avenue at Frankford Avenue. Kawasaki streetcars operated on Route 15 from June 18, 1995 to September 9, 1995, because Route 10 operated as a bus route due to a Philadelphia Water Department project at 48th Street and Lancaster Avenue. (*Kenneth C. Springirth photograph*)

The Richmond and Cumberland Streets Loop in the Olde Richmond neighborhood of Philadelphia on July 14, 1995 is the location of Route 15 Kawasaki car No. 9025 (delivered July 23, 1982 and placed in revenue service on July 30, 1982). From June 18, 1995 to September 9, 1995, Kawasaki cars operated on Route 15 from Cumberland Loop to 63rd and Malvern Avenue using Route 10 trackage on 63rd Street with buses operating from Cumberland Loop to the Route 15 terminus at Richmond and Westmoreland Streets. (*Kenneth C. Springirth photograph*)

On September 4, 2005, (the first day of service on the restored Route 15 streetcar) PCCII car No. 2335 is at the Richmond and Westmoreland Street Loop. This was originally car No. 2158 that was delivered to PTC in June 1948, overhauled by SEPTA in June 1984, and completely rebuilt by Brookville Equipment Corporation into PCCII car No. 2335. Each of the eighteen PCCII cars (Nos 2320–2337) is a completely new car built within an old PCC car shell that has been rebuilt. (*Kenneth C. Springirth photograph*)

Route 15 PCCII car No. 2326 is on Richmond Street at Allegheny Avenue on September 16, 2005. The longevity of this car began with St. Louis car Company delivering car No. 2197 to PTC in July 1948. It was overhauled by SEPTA in April 1985 and was completely rebuilt by the Brookville Equipment Corporation and returned to SEPTA on April 28, 2004 as car No. 2326. (*Kenneth C. Springirth photograph*)

Passing by a Piekarnia (Polish) bakery on Richmond Street near Clearfield Street on September 16, 2005, Route 15 car No. 2320 (one of eighteen PCC cars completely rebuilt by the Brookville Equipment Corporation and renumbered into PCCII cars [Nos 2320–2337]) is on its eastbound trip to 63rd Street and Girard Avenue, which from the Richmond and Westmoreland Loop could take from thirty-four minutes in the early morning hours to fifty-two minutes at rush hour. (*Kenneth C. Springirth photograph*)

Richmond Street at Indiana Avenue is the location of eastbound Route 15 car No. 2328 on October 5, 2005, with the Interstate 95 overpass clearly visible one block away. This originally was car No. 2726 delivered to PTC in March 1947. It was overhauled by SEPTA in April 1987 and sent to Brookville Equipment Corporation on September 23, 2002 where it was completely rebuilt and returned to SEPTA on May 27, 2004 as PCCII car 2328. (*Kenneth C. Springirth photograph*)

Westbound Route 15 PCCII car No. 2335 is on the Schuylkill River Bridge on November 11, 2005. This is the third Girard Avenue Bridge and was built between 1969 and 1972. In the background is the stone arch Pennsylvania Railroad Connecting Railway Bridge completed in 1915 (replacing an 1867 iron truss bridge) that in January 2019 was used by Amtrak's Northeast Corridor trains, SEPTA's Trenton and Chestnut Hill West Lines, and New Jersey Transit's Atlantic City Line. (*Kenneth C. Springirth photograph*)

On January 1, 2006, Route 15 PCCII car No. 2337 is passing by Girard College on Girard Avenue at Corinthian Avenue in the Francisville neighborhood of North Philadelphia. Originally, this was PCC car No. 2783 that was received by PTC in April 1947, and SEPTA overhauled it in February 1986. The car was completely rebuilt by Brookville Equipment Corporation and returned to SEPTA on November 23, 2004 as PCCII car No. 2337. (*Kenneth C. Springirth photograph*)

Poplar Street at 25th Street in the Brewerytown neighborhood of North Philadelphia is the location of eastbound Route 15 PCCII car No. 2320 on January 1, 2006. This 46.5-foot-long by 8.42-foot-wide car has seating for forty-six passengers and a capacity for up to seventy seated and standing passengers. The track that is turning north on the right side of the picture is the westbound track. (*Kenneth C. Springirth photograph*)

On January 2, 2006, Route 15 PCCII car No. 2328 is eastbound on Girard Avenue at 8th Street in the East Poplar neighborhood of Lower North Philadelphia. The bridge in the background was the former Reading Railroad commuter line that in January 2019 was used by SEPTA's Warminster, West Trenton, Lansdale - Doylestown, Chestnut Hill East, Manayunk - Norristown, and Fox Chase Lines. (*Kenneth C. Springirth photograph*)

Girard Avenue at 6th Street is the location of westbound Route 15 PCCII car No. 2333 on January 2, 2006. The car was originally No. 2746 delivered to PTC in March 1947 and overhauled by SEPTA in March 1985. It was completely rebuilt by Brookville Equipment Corporation and delivered to SEPTA as car No. 2333 on September 14, 2004. (*Kenneth C. Springirth photograph*)

On February 4, 2006, Route 15 PCCII car No. 2336 is on Girard Avenue at 59th Street in the Carroll Park neighborhood of West Philadelphia. Originally, this was car No. 2790 that PTC received from St. Louis Car Company in April 1947. It was overhauled by SEPTA in March 1983 and completely rebuilt by Brookville Equipment Corporation with delivery to SEPTA on November 4, 2004 as car No. 2336. (*Kenneth C. Springirth photograph*)

Under Interstate 95, PCCII car No. 2337 is turning from the Cumberland Loop to Richmond Street on October 15, 2006. Before the building of the Frankford and Delaware Avenue Loop, this was an important cutback loop when a truck would get stuck east of this location under a bridge blocking the streetcar line. (*Kenneth C. Springirth photograph*)

On October 15, 2006, Kawasaki car No. 9000 on a rail excursion is making a photo stop on Girard Avenue at the shelter that serves the Philadelphia Zoo. Route 15 is the only line that uses PCCII cars. Kawasaki cars normally operate on subway surface Routes 10, 11, 13, 34, and 36. (*Kenneth C. Springirth photograph*)

Route 15 PCCII car No. 2330 is on Girard Avenue at Front Street in this overhead view from the Girard Avenue Station of the Market - Frankford Elevated Line on May 19, 2007. This originally was car No. 2730 received by PTC in March 1947. It was overhauled by SEPTA in May 1986. Brookville Equipment Corporation completely rebuilt this car and delivered it to SEPTA as car No. 2330 in July 14, 2004. (*Kenneth C. Springirth photograph*)

Girard Avenue at Front Street at the Girard Avenue Station of the Market - Frankford Elevated Line in the Northern Liberties neighborhood of Philadelphia is the location of Route 15 PCCII car No. 2326 on May 19, 2007. This station is only a nine-minute ride by the Market - Frankford Line to the 15th Street Station in Center City Philadelphia, which has made this neighborhood one of Philadelphia's most desirable development districts attracting young professionals, students, artists, and design professionals. (*Kenneth C. Springirth photograph*)

On October 15, 2008, Kawasaki car No. 9000 is on Girard Avenue at Corinthian Ave on a rail excursion on Route 15. Behind the streetcar is Girard College named for Stephen Girard who directed the City of Philadelphia to use his money to build a school for poor orphaned boys. The school opened on January 1, 1848. Today, the school is a first through twelfth grade, five day a week boarding school for children from families with limited financial resources. (*Kenneth C. Springirth photograph*)

Girard Avenue at Alden Street in the Carroll Park neighborhood of West Philadelphia is the location of Route 15 PCCII car No. 2334 on June 7, 2009. This restored streetcar has a heating ventilation, air-conditioning (HVAC) system, an AC propulsion system, and Brookville soft-ride trucks to improve ride quality. (*Kenneth C. Springirth photograph*)

On August 6, 2010, Route 15 PCCII car No. 2336 is on Girard Avenue at Interstate 95. The car has an ADA lift installation, interior passenger stop request system, a public address system, original-style Art-Deco lighting, and stainless steel seating. (*Kenneth C. Springirth photograph*)

Girard Avenue at Crease Street is the location of eastbound Route 15 PCCII car No. 2329 on August 6, 2010. Originally, this was car No. 2182 that PTC received from St. Louis Car Company in July 1948. It was overhauled by SEPTA in April 1986. Brookville Equipment Corporation completely rebuilt this car with delivery to SEPTA as car No. 2329 on June 23, 2004. (*Kenneth C. Springirth photograph*)

On a rainy October 1, 2010, Route 15 PCCII car No. 2331 is waiting for departure time at the Richmond Street and Westmoreland Street Loop. This was originally car No. 2758 delivered to PTC in March 1947. It was overhauled by SEPTA in November 1985, and was completely rebuilt by Brookville Equipment Corporation and delivered to SEPTA on July 30, 2004. (*Kenneth C. Springirth photograph*)

In preparation for the reconstruction of Interstate 95, new switches have been installed for Route 15 at Girard and Frankford Avenues and concrete has been poured in this October 28, 2010 work scene. New track was placed in Frankford Avenue from Girard Avenue to the new loop at Frankford and Delaware Avenues. The first test run on this new trackage was on May 9, 2011 using Kawasaki car No. 9103 and PCCII car No. 2336. (*Kenneth C. Springirth photograph*)

On April 29, 2012, Route 15 PCC II car No. 2334 is on Girard Avenue at Howard Street, which is two blocks west of Front Street where the Girard Avenue Station of the Market - Frankford Elevated Line is located. The streetcar line has played a role in the redeveloping of Girard Avenue as evidenced by the refurbishing of a building to the right of the streetcar. (*Kenneth C. Springirth photograph*)

Westbound Route 15 PCCII car No. 2334 is on Girard Avenue at 2nd Street on April 29, 2012 passing the Hyperion Bank Building that was built in 1871 as the Eighth National Bank of Philadelphia. The building was designed in the Italianate/Second-Empire style, with its distinctive roof supported by pairs of large carved wood brackets, wide projecting cornices, windows with limestone arches arranged in pairs, and ornamented entranceways. Over a period of time, the building fell into disrepair and was abandoned. It was restored by Hyperion Bank. (*Kenneth C. Springirth photograph*)

On June 11, 2018, Route 15 PCCII car No. 2327 is turning from Girard Avenue onto Frankford Avenue passing by the historic former Kensington National Bank founded in 1826 and is now a Wells Fargo Bank. The streetcar was originally No. 2747 delivered to PTC in March 1947. It was overhauled by SEPTA in June 1985 and was completely rebuilt by Brookville Equipment Corporation with delivery to SEPTA on May 13, 2004 as car No. 2327. (*Kenneth C. Springirth photograph*)

At the Northern Liberties Loop on Frankford at Delaware Avenues on June 11, 2018, Route 15 PCCII car No. 2327 is awaiting departure time. This has been the eastern terminus of Route 15 during the reconstruction of Interstate 95. In fiscal year 2017, Route 15 had an average weekday ridership of 8,249 and required twelve cars during rush hours. (*Kenneth C. Springirth photograph*)

On June 11, 2018, Route 15 PCCII car No. 2322 has left the Northern Liberties Loop and is on Frankford Avenue at Allen Street. Originally, this was car No. 2770 delivered to PTC in April 1947 and was overhauled by SEPTA in May 1985. It was completely rebuilt by Brookville Equipment Corporation and delivered to SEPTA as car No. 2322 on January 29, 2004. (*Kenneth C. Springirth photograph*)

PTC Route 23 PCC car No. 2742 is on Germantown Avenue crossing the streetcar tracks used by Routes 53 and 56 on Erie Avenue in the Tioga neighborhood of North Philadelphia on May 30, 1961. This car was delivered in March 1947 and scrapped in 1977. (*Kenneth C. Springirth photograph*)

Northbound for the cutback Loop at Mermaid Lane, Route 23 PCC car No. 2742 (five years older than on the top picture of this page) is on Germantown Avenue at the Germantown Depot on July 20, 1966. The People's Passenger Railway began electric streetcar service from Dauphin Street to Westfield Street (location of the Germantown Depot) on August 12, 1894. People's Traction Company took over on February 3, 1895. It became part of the Union Traction Company on July 1, 1896 which merged into PRT on May 1, 1902, and became PTC on January 1, 1940. (*Kenneth C. Springirth photograph*)

On a snow-covered February 22, 1969, SEPTA Route 23 PCC car No. 2786 is at the Germantown Depot waiting for the next assignment. This car was delivered to PTC in April 1947 and was scrapped in 1977. In September 1976, Route 23 streetcars were moved from Germantown Depot to Luzerne Depot. (*Kenneth C. Springirth photograph*)

On April 27, 1968 on 11th Street at Walnut Street, northbound Route 23 PCC car No. 2732 with the red slash over '23' tells the rider the car will not cover the complete route, and in this case will go to the Mermaid Lane Loop. Delivered to PTC in March 1947, this car was overhauled by SEPTA in February 1985. In 2005, it was sold to Glen Echo Park, and in 2012 sold to General Machine & Tool Company of Cheverly, Maryland. (*Kenneth C. Springirth photograph*)

Southbound Route 23 PCC car No. 2728 is on 12th Street having just crossed Snyder Avenue (double wires are for trackless trolley Route 79) in the East Passyunk neighborhood of South Philadelphia on December 26, 1970. Delivered to PTC in March 1947, this was the 100th car to be overhauled by SEPTA in May 1986, went to the Baltimore Streetcar Museum in 2005, was placed on display in downtown Lancaster, Pennsylvania, and in 2012 went to Warren Rail Car in Warren, Pennsylvania. (*Kenneth C. Springirth photograph*)

On December 26, 1970, PCC car No. 2740 is at the 10th and Bigler Streets terminus of Route 23 in the Stadium District of South Philadelphia. PTC received this car in March 1947. It was overhauled by SEPTA in October 1983, and was acquired by the National Museum of Transportation 2933 Barrett Station Road, St. Louis, Missouri, in October 1994. (*Kenneth C. Springirth photograph*)

Route 23 PCC car No. 2769 is southbound on 12th Street at Ritner Street in the Lower Moyamensing neighborhood of South Philadelphia on December 26, 1970. This car was delivered to PTC in April 1947 and was scrapped by SEPTA in 1977. (*Kenneth C. Springirth photograph*)

The Germantown Avenue and Bethlehem Pike northern terminus of Route 23 in the Chestnut Hill neighborhood of Philadelphia is the location of PCC car No. 2261 in 1979. This was originally Kansas City Public Service Company car No. 786 built by St. Louis Car Company in 1946. It was sold to SEPTA in 1955 where it was refurbished and renumbered 2261. In November 1985, it was sold to the Vintage Electric Streetcar Company of Windber, Pennsylvania. (*Clifford R. Scholes collection*)

In 1979, Route 23 PCC car No. 2278 is on Germantown Avenue at Hunting Park Avenue in the Nicetown neighborhood of North Philadelphia. This originally was car No. 781 delivered to Kansas City Public Service Company in 1946. It was sold to Philadelphia Transportation Company and refurbished in July 1955. Under the Southeastern Pennsylvania Transportation Authority it received a Bicentennial paint scheme and lettered New Hampshire. In April 2005, the car was stored outside at McDale Park in Lackawanna County, Pennsylvania. (*Clifford R. Scholes collection*)

The Chestnut Hill Loop is the location of Route 23 PCC car No. 2669 on April 10, 1982. This was one of forty-seven cars (Nos 2634–2680) delivered by St. Louis Car Company in 1942. Powered by four Westinghouse type 1432D motors, each car weighed 35,860 pounds. This car was repainted cream and green without the narrow maroon band just below the windows. (*Kenneth C. Springirth photograph*)

In 1983, PCC cars Nos 2186 (delivered to PTC in July 1948 and overhauled by SEPTA in November 1983) and 2263 (delivered to Kansas City Public Service Company as car No. 768 in 1946, sold to PTC in 1955, and scrapped in November 1983) are at the Chestnut Hill Loop on an operator layover. Both of these cars were built by St. Louis Car Company. Car 2186 was acquired by the National Museum of Transport of St. Louis, Missouri, in November 1994. It was sold to the San Diego Vintage Trolley and renumbered 532. (*Clifford R. Scholes collection*)

Northbound Route 23 PCC car No. 2111 is on East Passyunk Avenue in the Passyunk Square neighborhood of South Philadelphia in August 1985. This car was delivered to PTC in August 1948, overhauled by SEPTA in October 1982, and sold to Brookville Equipment Corporation in 2004. (*Kenneth C. Springirth photograph*)

In August 1985, southbound Route 23 PCC car No. 2092 is on 12th Street crossing Market Street in Center City Philadelphia. St. Louis Car Company delivered this car to PTC in August 1948. It was overhauled by SEPTA in March 1982 and was sold to Brookville Equipment Corporation in 2004. (*Kenneth C. Springirth photograph*)

On 11th Street at Snyder Avenue Route 23, PCC car No. 2751 is northbound for the long trip to Chestnut Hill Loop in August 1985. This car was delivered to PTC in March 1947, overhauled by SEPTA in September 1984, and sold to the New Orleans Regional Transit Authority in December 1994. The car body was scrapped in March 1997. (*Kenneth C. Springirth photograph*)

Northbound Route 23 PCC car No. 2741 is on 11th Street at Bainbridge Street in the Hawthorne neighborhood of South Philadelphia on August 8, 1985. PTC received this car in March 1947. It was overhauled by SEPTA in May 1984, and completely rebuilt by Brookville Equipment Corporation with delivery to SEPTA as car No. 2325 on April 14, 2004. (*Kenneth C. Springirth photograph*)

In the bright sunshine of June 28, 1986, PCC car No. 2123 is at the Route 23 southern terminus at 10th and Bigler Streets waiting for departure time. This car was delivered to PTC in August 1948. SEPTA overhauled this car in October 1981, and sold it to the San Francisco Municipal Railway (SFMR) where it was restored by Morrison-Knudsen Corporation in 1993 and renumbered 1051. In 2016, it was rebuilt by Brookville Equipment Corporation and painted in the SFMR simplified green and cream paint scheme. (*Kenneth C. Springirth photograph*)

Southbound Route 23 PCC car No. 2777 is on 12th Street at Snyder Avenue on June 28, 1986. PTC received this car in April 1947. It was overhauled by SEPTA in February 1986, and was completely rebuilt by Brookville Equipment Corporation, renumbered 2324, and delivered to SEPTA on March 26, 2004. (*Kenneth C. Springirth photograph*)

Westbound on Oregon Avenue, Route 23 PCC car No. 2182 is about to turn north on 11th Street for a neat way to see the many neighborhoods of Philadelphia on the trip to Chestnut Hill on June 28, 1986. This car was completely rebuilt by Brookville Equipment Corporation and renumbered 2329 with delivery to SEPTA on June 23, 2004. This was one of eighteen PCC cars completely rebuilt for Route 15. (*Kenneth C. Springirth photograph*)

On November 28, 1986, northbound Route 23 PCC car No. 2170 is at a passenger top on 11th Street at Market Street in Center City Philadelphia. Delivered to PTC in July 1948, this car was overhauled by SEPTA in October 1986, and was sold to Brookville Equipment Corporation for parts on February 27, 2003. (*Kenneth C. Springirth photograph*)

The Route 23 southern terminus on 10th at Bigler Streets is the location of PCC car No. 2788 in February 1987. PTC received this car in April 1947. It was overhauled by SEPTA in May 1984 and sold to the New Orleans Regional Transit Authority in December 1994 where the car body was scrapped in March 1997. (*Kenneth C. Springirth photograph*)

On April 28, 1987, Route 23 PCC car No. 2748 is leaving picturesque Mermaid Lane Loop without the route number showing on the front destination sign. St. Louis Car Company delivered this car to PTC in March 1947, and it was overhauled by SEPTA in March 1986. The car was scrapped at Midvale Depot in January 2006. (*Kenneth C. Springirth photograph*)

On October 3, 1992, repainted PCC car No. 2168 is on Germantown Avenue at Southampton Avenue on a "Chestnut Hill Trolley" trip. Following the conversion of Route 23 to bus operation on February 21, 1992, SEPTA began on September 13, 1992 and ending on June 15, 1996 (because of budget cuts) a weekends only Chestnut Hill Trolley on Germantown Avenue from the Chestnut Hill Loop to the Germantown Depot. (*Kenneth C. Springirth photograph*)

Kawasaki car No. 9045 is decorated for Halloween and is on former Route 23 on 11th at Bainbridge Streets on October 22, 1995. This was part of Trolley Fest held by SEPTA on October 21 and 22, 1995, which included tours of the Elmwood Depot and Woodland Shop, plus trolley parades, and a variety of special trolley excursions. On May 21, 1995, SEPTA began Route 51 the "Welcome Line" on a portion of Route 23 on 11th and 12th Streets between Noble and Bainbridge Streets with certain trips extended south to Snyder Avenue. (*Kenneth C. Springirth photograph*)

On November 11, 1998, two SEPTA "Holiday Trolleys" PCC cars (Nos 2168 and 2799) are on 12th Street at Arch Street in Center City Philadelphia ready for the next run. The "Holiday Trolley" operated on the same portion of Route 23 as the "Welcome Line" with service about every twenty minutes. (*Kenneth C. Springirth photograph*)

5

Routes 34 and 36

Route 34 operated from the Juniper Street Subway Station to its western terminus at 62nd and Baltimore Avenue beginning August 24, 1908. In 1912, the western terminus became 61st and Baltimore Avenue, which was a connecting point with the Delaware County & Philadelphia Electric Railway Company streetcar line to Media. This later became Southern Pennsylvania Traction Company, which discontinued streetcar service from Philadelphia to Media on August 2, 1930. From the subway surface tunnel under Market Street, Route 34 exits the subway at the 40th Street Portal and heads west on Baltimore Avenue to the loop at 61st Street. This is the shortest of SEPTA's five Subway Surface streetcar lines.

Route 36 began on March 15, 1904 operating from Island and Elmwood to 49th and Woodland. It was extended east to 9th and Arch Streets on May 18, 1905. Effective November 6, 1955, Route 37 was merged into Route 36 operating from Juniper Street Subway Station to Westinghouse Loop. The line was cutback numerous times as follows: 94th and Eastwick on September 9, 1956, 88th and Eastwick on August 15, 1962, 84th and Eastwick on January 5, 1966, and to 80th and Eastwick on April 26, 1975. From the subway surface tunnel under Market Street, Route 36, the longest of SEPTA's five subway surface lines, exits the subway at the 40th Street Portal, runs southwest on Woodland Avenue with Route 11, turns south on 49th Street, turns heading southwest on Grays Avenue, southwest on Lindbergh Boulevard, and makes a slight turn to go southwest on Elmwood Avenue. Passing the Elmwood Depot, Route 36 turns south on Island Road to the 80th and Eastwick Avenue terminus.

The 61st Street and Baltimore Avenue Loop is the location of Route 34 PCC car No. 2039 in 1968. This was one of ten PCC cars (Nos 2031–2040) built by St. Louis Car Company and delivered in February 1941. Powered by four Westinghouse type 1432D motors, each car weighed 35,890 pounds and seated fifty passengers. The car was destroyed in the Woodland Depot fire of October 23, 1975. (*Clifford R. Scholes collection*)

In 1982, Route 34 Kawasaki car No. 9054 (delivered on September 28, 1981 and placed in revenue service on October 6, 1981) is at the Baltimore Avenue and 61st Street Loop in the Angora neighborhood of Southwest Philadelphia. Route 34 is the shortest of SEPTA's five subway surface lines. (*Clifford R. Scholes collection*)

On October 15, 2006, Route 34 Kawasaki car No. 9000 is on the diversion trackage on 42nd Street at Osage Avenue in the Spruce Hill neighborhood of West Philadelphia. This is part of the West Philadelphia Streetcar Suburb Historic District. The conversion of horsecar lines to electric streetcar operation reduced travel time to downtown Philadelphia and developers built vast housing tracts in this once rural area. (*Kenneth C. Springirth photograph*)

With the streetcar subway closed for a planned maintenance blitz, riders in the afternoon rush hour of August 15, 2014 are on 40th at Market Streets boarding Route 34 Kawasaki car No. 9080 (delivered October 13, 1982 and placed in revenue service on November 19, 1982). This is also the location of the 40th Street subway station of the Market - Frankford Line for riders to reach Center City Philadelphia. (*Kenneth C. Springirth photograph*)

On June 11, 2018, eastbound Route 34 Kawasaki car No. 9091 (delivered November 2, 1981 and placed in revenue service on November 17, 1981) is on Baltimore Avenue at 42nd Street. Route 34 service levels have been fairly stable except for Sunday where the frequency on August 26, 2018 was every twenty minutes compared to a fourteen-minute frequency on December 22, 1963. (*Kenneth C. Springirth photograph*)

With a mural of car No. 1775—one of 772 cars (Nos 1585–2356) built by J. G. Brill Company between 1899 and 1906 for the Union Traction Company and its successor the Philadelphia Rapid Transit Company—painted on the side of a building, on the right is SEPTA Route 34 Kawasaki car No. 9075 (delivered July 23, 1982 and placed in revenue service on July 29, 1982) handling an east bound trip on Baltimore Avenue at 45th Street on June 11, 2018. (*Kenneth C. Springirth photograph*)

On June 11, 2018, Baltimore Avenue at 42nd Street with the Covenant Community Church on the left is the location of Route 34 Kawasaki car No. 9049 (delivered September 21, 1981 and placed in revenue service on September 30, 1981). In 2017, the 4.9-mile Route 34 had an average weekday ridership of 12,562 and required twenty-two cars during rush hours. (*Kenneth C. Springirth photograph*)

The Elmwood Avenue and Island Road Loop in the Elmwood neighborhood in Southwest Philadelphia is the location of Route 36 PTC car No. 8350 in 1948. This car was delivered by J. G. Brill Company to PTC in December 1923, remodeled in May 1941, and scrapped in January 1956. (*Clifford R. Scholes collection*)

In 1956, Route 36 PCC car No. 2077 (delivered in March 1941 and scrapped in 1977) is at the Westinghouse Loop located on Powhattan and Jansen Avenues in Lester west of Philadelphia in Delaware County. Route 37 was combined into Route 36 on November 6, 1955 and operated from Juniper Street Subway Station to Lester. Route 36 was cutback to 94th Street and Eastwick Avenue on September 9, 1956. (*Clifford R. Scholes collection*)

The 94th and Eastwick Loop in the Eastwick Neighborhood of Southwest Philadelphia is the location of Route 36 car No. 2508 on April 24, 1960. This car was delivered to PTC in December 1940 and was scrapped in 1981. (*Kenneth C. Springirth photograph*)

Route 36 PCC car No. 2625 is on the private right of way eastbound to Center City Philadelphia on March 18, 1962. St. Louis Car Company delivered this car to PTC in June 1942, and it was scrapped in 1978 providing about thirty-six years of service. Route 36 service levels have been fairly stable except for Sunday where the frequency on August 26, 2018 was every twenty minutes compared to a twelve-minute frequency on January 21, 1968. (*Kenneth C. Springirth photograph*)

In 1970, Route 36 PCC car No. 2777 has emerged from the 40th Street Subway Portal. PTC received this car in April 1947. It was overhauled by SEPTA in 1986, and was completely rebuilt by Brookville Equipment Corporation and delivered to SEPTA as No. 2324 on March 26, 2004. (*Clifford R. Scholes collection*)

Elmwood Avenue at Island Avenue on Route 36 is the location of SEPTA PCC car No. 2124 for a photo stop on an Electric Railroaders' Association Inc. Convention rail excursion on July 7, 1973. (*Kenneth C. Springirth photograph*)

In 1982, Route 36 Kawasaki car No. 9002 is leaving the 80th Street and Eastwick Avenue Loop. In the background on the right side of the picture is a rear view of PCC car No. 2054 on a rail excursion photo stop. (*Clifford R. Scholes collection*)

The Elmwood Depot on Elmwood and Island Avenues is under construction in this 1981 scene with a 2100 series PCC car in view. Opening on November 20, 1981, Elmwood Depot stores and maintains the Kawasaki cars for SEPTA Routes 11, 13, 34, and 36. Routes 10 and 15 streetcars are based at Callowhill Depot in West Philadelphia. (*Clifford R. Scholes collection*)

On January 5, 2000, PCC car No. 2799 painted in a Philadelphia Suburban Transportation Company paint scheme is at the 80th Street and Eastwick Avenue Loop for a photo stop on a rail excursion. Delivered to PTC in April 1947, the car was overhauled by SEPTA in August 1985, was on exhibit in downtown Lancaster, Pennsylvania, and in 2012 went to Warren Rail Car in Warren, Pennsylvania. (*Kenneth C. Springirth photograph*)

Southbound Route 36 Kawasaki car No. 9072 (delivered on March 29, 1982 and placed in revenue service on April 2, 1982) is on private right of way heading for the 80th Street and Eastwick Avenue Loop in 1988. Route 36 is the longest of the five subway surface routes and is the only streetcar route in the City of Philadelphia that operates on some private right of way. (*Clifford R. Scholes collection*)

On July 31, 1998, Route 36 Kawasaki car No. 9047 (delivered on September 18, 1981 and placed in revenue service on September 24, 1981) is on Elmwood Avenue near Island Avenue. On the left side of the picture is the access trackage from Elmwood Depot. (*Kenneth C. Springirth photograph*)

PCC II car No. 2336 is at the Elmwood Depot for a rail excursion on February 5, 2005. The PCC II cars were housed at Callowhill Depot for Route 15, which returned to streetcar service on September 4, 2005. In advance of that return, they were used on various rail excursion charter trips. (*Kenneth C. Springirth photograph*)

PCCII car No. 2331 and PCC car No. 2168 are at Elmwood Depot on February 5, 2005. On the left, three Kawasaki cars are in the lineup of cars. Buses replaced streetcars on Route 15 on September 13, 1992; however, SEPTA later agreed to restore Route 15 streetcar service. Brookville Equipment Corporation completely rebuilt eighteen PCC cars into PCCII cars for Route 15, which resumed streetcar service on September 4, 2005. (*Kenneth C. Springirth photograph*)

At the 80th Street and Eastwick Avenue Loop on February 5, 2005, are Route 36 Kawasaki car No. 9040 in regular service and PCCII car No. 2337 on a rail excursion with the destination sign showing 37 Westinghouse, which once was a legitimate destination. Route 37 was combined into Route 36 on November 6, 1955, with Route 36 operating from Juniper Street Subway Station to Westinghouse loop until September 9, 1956 when it was cutback from Westinghouse Loop to 94th Street and Eastwick Avenue. (*Kenneth C. Springirth photograph*)

PCC overhead line car No. 2194, in a bright orange to improve visibility for service at night, and behind it PCCII car No. 2334 are at the Elmwood Depot on a Friends of Philadelphia Trolleys rail excursion on June 7, 2009. (*Kenneth C. Springirth photograph*)

On June 11, 2018, Kawasaki car No. 9039 with the Route 36 destination of 73rd – Elmwood and special front window sign "Student Driver No Passengers" is turning to avoid going into the 40th Street Subway Portal. In fiscal year 2017, the 7-mile Route 36 had an average weekday ridership of 12,856 and required twenty-one cars during rush hours. (*Kenneth C. Springirth photograph*)

6

Routes 47 and 50

Route 47 operated from Front and Arch Streets to 15th Street and Glenwood Avenue on July 6, 1910. After many routing changes on November 4, 1951, Route 47 operated from 5th Street and Godfrey Avenue to 8th and Wolf Streets. Leaving 5th and Godfrey the line headed south on 5th Street, southwest on Rising Sun Avenue, south on 6th Street, west on Susquehanna Avenue, and south on 8th Street to the southern terminus at 8th and Wolf Streets. Northbound trips went west on Wolf Street to 9th Street, north on 9th Street, east on Spring Garden, north on 7th Street, northwest on Germantown Avenue, north on 8th Street, east on Allegheny Avenue, and north on 5th Street to the northern terminus at Godfrey Avenue. Effective July 9, 1967, Philadelphia Transportation Company (PTC) converted the southern end of the line to bus operation due to construction of the new Lindenwold Line. Route 47 streetcars operated from 5th and Godfrey Avenue south to 8th and Spring Garden Streets. The Southeastern Pennsylvania Transportation Authority (SEPTA) purchased PTC on September 30, 1968, and converted the northern end of Route 47 to bus operation on June 15, 1969.

Route 50 first operated from 5th and Jackson to the Fox Chase Loop at Oxford Avenue and Loney Street on April 13, 1913. Effective September 28, 1913, Route 50 operated from 4th and Ritner Streets to the Fox Chase Loop. On April 5, 1953, Route 50 operated from 6th and Oregon to the Rising Sun and Knorr Loop. Southbound cars went southwest on Rising Sun Avenue, west on Wyoming Avenue, south on 5th Street, southwest on Rising Sun Avenue, south on 6th Street, east on Oxford Street, south on 4th Street, west on Snyder Avenue, south on 6th Street, and west on Oregon Avenue to 7th Street. Peak period trips operated south to 4th and Ritner Streets. Northbound cars from Oregon Avenue went north on 7th Street, east on Snyder Avenue, north on 5th Street, east on Wyoming Avenue, and northeast on Rising Sun Avenue to Olney Avenue with peak period trips continuing north to Knorr Street. Route 50 was converted to bus operation on October 12, 1980.

Route 47 PCC car No. 2154 (delivered to PTC in June 1948 and scrapped in March 1974) is on 8th Street at Green Street in the Poplar neighborhood of Lower North Philadelphia on April 27, 1968. On July 9, 1967, PTC converted the southern end of Route 47 to bus operation due to construction of the new Lindenwold Line. The northern end of Route 47 streetcars operated from 5th and Godfrey Avenue as far south as 8th and Spring Garden Street. (*Kenneth C. Springirth photograph*)

On May 31, 1968, Route 47 PCC car No. 2727 (delivered to PTC in March 1947 and scrapped in 1982) is on 5th Street at Ruscomb Street, which is about two blocks north of Roosevelt Boulevard in the Olney neighborhood of Philadelphia. (*Kenneth C. Springirth photograph*)

PTC Route 50 PCC car No. 2148 is turning from 6th Street to Oregon Avenue in the Whitman neighborhood of South Philadelphia on March 4, 1962. This car was delivered in June 1948 and was destroyed in the Woodland Depot fire of October 23, 1975. (*Kenneth C. Springirth photograph*)

On April 20, 1962, Route 50 PCC car No. 2282 is turning from Rising Sun Avenue into the Knorr Street Loop in the Lawndale neighborhood of Northeast Philadelphia. This car was originally No. 738 built in 1946 by St. Louis Car Company for Kansas City Public Service Company. It was sold and delivered to PTC in June 1955. The car was scrapped in June 1972. (*Kenneth C. Springirth photograph*)

Northbound on 5th Street crossing Market Street with Old City Hall on the left and the bell tower of Independence Hall in the center of the Independence Hall complex in Center City Philadelphia on July 21, 1962, Route 50 PCC car No. 2264 amid vintage automobiles has a long journey to its Rising Sun and Olney destination. This car was originally Kansas City Public Service car No. 742 that was acquired by PTC in April 1955. It was scrapped in 1976. (*Kenneth C. Springirth photograph*)

On April 27, 1968, PTC Route 50 car No. 2145 is northbound on 5th Street at Erie Avenue in the Hunting Park neighborhood of North Philadelphia. PTC received this car from St. Louis Car Company in June 1948, and it was destroyed in the Woodland Depot of fire of October 23, 1975. (*Kenneth C. Springirth photograph*)

Route 50 PCC car No. 2283 is on Wyoming Avenue at 3rd Street in the Feltonville neighborhood of North Philadelphia on May 31, 1968. The double overhead wire was used by the Route 75 trackless trolley. This car was originally Kansas City Public Service Company car No. 734, which was acquired by PTC in June 1955. It was sold in November 1985 to Vintage Electric Streetcar Company of Windber, Pennsylvania. (*Kenneth C. Springirth photograph*)

In the densely populated Queen Village neighborhood, just south of Center City Philadelphia, Route 50 PCC car No. 2513 is northbound on 5th Street at Catharine Street on December 26, 1970. Delivered to PTC in December 1940, this car was scrapped over forty-one years later in March 1982. (*Kenneth C. Springirth photograph*)

On 4th Street at Christian Street in Philadelphia's Queen Village neighborhood on December 26, 1970, Route 50 PCC car No. 2077 is southbound for 6th Street and Oregon Avenue. During Monday to Friday morning and afternoon peak periods, certain trips terminated at 4th and Ritner Streets. Compare this picture with the same car fourteen years earlier on the bottom of page 77 to see the effect of a busy urban environment. (*Kenneth C. Springirth photograph*)

On May 28, 1976, Route 50 PCC car No. 2278 is southbound on 6th Street at Allegheny Avenue in the Franklinville neighborhood of North Philadelphia. This was one of twenty-seven cars of the series 2251–2290 to receive the special Bicentennial paint scheme. Cars Nos 2270 and 2278 were lettered "New Hampshire." (*Kenneth C. Springirth photograph*)

Southbound on 4th Street at Walnut Street on August 14, 1976, Route 50 PCC car No. 2261 in a Bicentennial paint scheme and lettered "Virginia" is passing by the Independence National Historical Park with a view on the left of the Second Bank of the United States designed by architect William Strickland, with its Greek Revival architectural style characterized by columns along the front and a gabled roof with two sloping sides that meet at a ridge line on the roof. (*Kenneth C. Springirth photograph*)

7

Routes 53 and 56

Route 53 on April 15, 1904 operated from Wayne Avenue and Carpenter Lane to 13th and Locust Streets. Effective September 7, 1930, it operated from Wayne and Carpenter Loop to the 10th and Luzerne Streets Loop. On August 14, 1938, Route 53 was the first line in Philadelphia to have PCC cars. Effective September 11, 1938 on Sundays, it was consolidated with Route 75 to operate to Richmond and Lefevre Streets until April 19, 1948 when Route 75 was converted to trackless trolley operation. Leaving the Carpenter Lane and Wayne Avenue Loop, Route 53 headed southeast on Wayne Avenue, southeast on Clarissa Street, south on 18th Street, southeast on Pulaski Avenue, east on Erie Avenue (which also carried Route 56 streetcars), and north on 10th Street to the Luzerne Depot Loop. Route 53 was converted to bus operation on May 16, 1985.

Route 56 operated from 23rd and Venango Streets to Frankford and Erie Avenues on December 4, 1927. It was extended to Torresdale and Cottman Avenues on September 2, 1928. This was the last major new streetcar track extension in Philadelphia. From the Venango Street Loop at West Hunting Park Avenue and 23rd Street, Route 56 headed east on Erie Avenue. From Pulaski Avenue to 10th Street, it was joined by streetcar Route 53. At the Erie – Torresdale Station of the Market -Frankford Elevated Line over Kensington Avenue, Erie Avenue becomes Torresdale Avenue, and the line continues east to the Torresdale and Cottman Avenues Loop. Route 56 was converted to bus operation on June 14, 1992.

Waiting for departure time is Route 53 PCC car No. 2548 (delivered in January 1941 and scrapped in October 1972) at the 10th Street and Luzerne Street Loop on July 21, 1962. Cars Nos 2001–2090 and 2501–2680 were air-electric PCC cars, with compressed air powering the brakes, doors, and windshield wipers. (*Kenneth C. Springirth photograph*)

In 1974, with a touch of snow on the ground, Route 53 PCC car No. 2059 is on Clarissa Street at Juniata Street in the Nicetown neighborhood of North Philadelphia. St. Louis Car Company delivered this car to PTC in February 1941, and it was scrapped in 1982. (*Clifford R. Scholes collection*)

The Carpenter Lane and Wayne Avenue Loop of Route 53 in the West Mount Airy neighborhood of Northwest Philadelphia is the location of SEPTA PCC car No. 2715 in 1981. The car was acquired by San Francisco Municipal Railway (SFMR) and was restored to represent Newark, New Jersey, by Morrison-Knudsen Corporation in 1993 as car No. 1060. After a November 2002 derailment, the car was repainted in the 1938 Philadelphia silver paint scheme with a cream window area and a blue stripe. (*Clifford R. Scholes collection*)

On April 27, 1968, PTC eastbound Route 56 PCC car No. 2558 (delivered in January 1941 and scrapped in 1981) is on Erie Avenue ready to cross the northbound Routes 47 and 50 streetcar trackage on 5th Street. (*Kenneth C. Springirth photograph*)

Torresdale Avenue with its classic row homes near Cottman Avenue in the Tacony neighborhood of Northeast Philadelphia is the location of Route 56 PCC car No. 2571 on April 27, 1968. This car was delivered by St. Louis Car Company to PTC in January 1941, and it was scrapped forty-two years later in 1983. (*Kenneth C. Springirth photograph*)

Route 56 PCC Car No. 2062 (delivered in February 1941 and scrapped in 1979) is at a passenger stop on Torresdale Avenue at Frankford Avenue in the Juniata Park neighborhood of Northeast Philadelphia in 1969. (*Clifford R. Scholes collection*)

In 1973, Route 56 PCC car No. 2290 is on Erie Avenue at 3rd Street. This car was originally Kansas City Public Service Car No. 752 built by St. Louis Car Company in 1946. It was sold to PTC in 1955 and sold to the Vintage Electric Streetcar Company in Windber, Pennsylvania, in November 1985. (*Clifford R. Scholes collection*)

PCC car No. 2749 is at the Route 56 Torresdale and Cottman Avenue Loop in the Tacony neighborhood of Northeast Philadelphia in 1982. PTC received this car in March 1947. It was overhauled by SEPTA in November 1983 and was sold to the New Orleans Regional Transit Authority in December 1994. (*Clifford R. Scholes collection*)

In 1984, Route 56 PCC car No. 2102 is westbound on Torresdale Avenue and Erie Avenue. Erie Avenue becomes Torresdale Avenue after crossing Kensington Avenue. This car was delivered to PTC in August 1948, overhauled by SEPTA in November 1982, and sold to Colorado Springs in July 1995. The building behind the streetcar was the Northeast Catholic High School that was purchased by the Marianna Bracetti Academy Charter School in September 2011. (*Clifford R. Scholes collection*)

On August 4, 1991, eastbound Route 56 PCC car No. 2185 is on Erie Avenue at Kensington Avenue which is a transfer point with the Erie - Torresdale Elevated Station of the Market - Frankford Line. PTC received this car in July 1985. It was overhauled by SEPTA in May 1984 and was acquired by the East Troy Electric Railroad of East Troy, Wisconsin, in November 1994. The car was purchased by John DeLamater and donated to Kenosha Area Transit of Kenosha, Wisconsin, in October 2011. (*Kenneth C. Springirth photograph*)

The Route 56 Torresdale and Cottman Avenues Loop is the location of PCC car No. 2777 waiting for departure time on August 4, 1991. This was rebuilt by Brookville Equipment Corporation and became No. 2324, joining seventeen other rebuilt cars for Route 15, which reopened as a streetcar line on September 4, 2005. (*Kenneth C. Springirth photograph*)

In 1986, Route 56 PCC car No. 2747 is on Erie Avenue at Front Street in the Feltonville neighborhood of North Philadelphia. This portion of Erie Avenue had signs indicating "Trolley Only" over the trolley tracks to keep motor vehicles off of the tracks. (*Clifford R. Scholes collection*)

On May 29, 1992, Route 56 PCC cars Nos 2186 (sold to National Museum of Transportation at St. Louis, Missouri, in November 1994 and in 2009 acquired by San Diego Vintage Trolley where it was renumbered 532) and 2730 (painted grey in the window area and rebuilt by Brookville Equipment Corporation with delivery to SEPTA on July 14, 2004 as No. 2330 for restored Route 15) are at the Torresdale and Cottman Avenues Loop. (*Kenneth C. Springirth photograph*)

Route 56 PCC car No. 2728 is westbound on Torresdale Avenue at Fraley Street in the Wissinoming neighborhood of Northeast Philadelphia on May 29, 1992. This car was on display in Lancaster, Pennsylvania, and in 2012 went to the Warren Rail Car in Warren, Pennsylvania. (*Kenneth C. Springirth photograph*)

On May 29, 1992, eastbound Route 56 PCC car No. 2777 is at a busy passenger stop on Torresdale Avenue at Kensington Avenue, with the Erie - Torresdale Elevated Station of the Market - Frankford Line behind the streetcar. This car was rebuilt by Brookville Equipment Corporation as No. 2324 for Route 15. (*Kenneth C. Springirth photograph*)

8

Routes 60 and 62

Route 60 operated from Richmond Street and Allegheny Avenue to 17th Street and Allegheny Avenue on November 1, 1906. It was extended to Ridge and Allegheny Avenues on September 10, 1939. Effective May 16, 1956, Route 60 operated from Richmond and Westmoreland Streets to 35th Street and Allegheny Avenue. From the 35th Street and Allegheny Avenue Loop, Route 60 operated east on Allegheny Avenue, northeast on Sedgley Avenue, turned southeast and shortly went back to Allegheny Avenue, and headed east to Kensington Avenue where it crossed under the Allegheny Station of the Market – Frankford Elevated Line. At Kensington Avenue, Allegheny Avenue turned southeast, and Route 60 was on Allegheny Avenue to Richmond Street where it turned northeast to the Richmond and Westmoreland Loop. On September 4, 1977, buses replaced Route 60 streetcars.

Route 62 opened on June 1, 1907 as an extension to Darby of the Philadelphia Rapid Transit Company's Chester Avenue line, which operated to Front and Chestnut Streets in Philadelphia. That line was cut back to Mount Moriah Cemetery, and Route 62 began operation on November 10, 1918 from 9th and Main Streets in Darby via Kingsessing Avenue to Mount Moriah Cemetery in Philadelphia. On June 19, 1927, Route 62 operated from 9th and Main Streets in Darby to Yeadon Loop on Chester at Callahan Avenues in Yeadon. Route 62, operating entirely in Delaware County, required one car for the five-minute trip each way. On January 24, 1971, Route 62 was replaced by extending certain Route 13 trips into Darby.

Route 60 PCC car No. 2263 is at the Richmond and Westmoreland Loop on June 13, 1971. This car was originally Kansas City Public Service Company car No. 768 built in 1946, was sold to PTC in 1955, and scrapped in November 1983. (*Kenneth C. Springirth photograph*)

In 1977, Route 60 PCC car No. 2290 is on Allegheny Avenue at Witte Street in the Richmond neighborhood of Philadelphia. Originally, PCC car No. 752 built in by St. Louis Car Company for the Kansas City Public Service Company, sold to PTC in 1955, renumbered 2290, and sold to Vintage Electric Streetcar Company in November 1985. (*Clifford R. Scholes collection*)

On the private right of way in Darby, Route 62 Nearside car No. 7451 is eastbound for Yeadon in 1952. This car was one of 500 cars (Nos 7000–7499) built by J. G. Brill Company in 1913. Powered by two General Electric type 201P motors, each car weighed 36,000 pounds. Delivered in October 1913, the car was scrapped in June 1955. (*Clifford R. Scholes collection*)

9

Work Cars

Philadelphia Rapid Transit Company (PRT) ordered fifty-six work cars from J. G. Brill Company in 1923, consisting of twenty-four double-truck plows, ten double-truck sweepers, eight double-truck dump motor cars, five double-truck trailers, six double-truck flat trailers, one double-truck surface crane, one double-truck elevated crane, and one double-truck line car. At its peak, PRT had 127 snow sweepers, and they were designated class C. Single truck snow sweepers had short brooms and small wing plows. Both brooms operated counterclockwise at the same time throwing the snow out in front of the car. Double-truck snow sweepers had full width brooms, and only the broom at the forward end of the car operated. Class D cars were used by the Electrical Department of PRT for the production of power plus maintenance of transmission lines and overhead wire. Beginning 1903, coal was hauled to power stations by wooden trolley coal cars. During 1908, two steel hopper trolley coal cars Nos 2624 and 2625 (later renumbered D-7 and D-8) were added. At its peak, PRT had forty roadway department cars, and they were designated class W. These cars were used for hauling roadway supplies. Class S sand cars peaked at thirty-nine cars and were used to spread sand over the rails to improve traction for the regular service cars.

The Frankford Depot is the location of sand car No. S-2 in 1953. This was built by J. G. Brill Company on April 8, 1913 and scrapped on March 6, 1958. This car, along with sand car No. S-1, were specially designed for the hauling of sand with steel tanks holding 15 cubic yards of sand. Each car had a bottom spout to fill the underground storage hoppers by gravity. For a location that did not have an underground storage hopper, the car bottom had a connection for a 4-inch hose to force the sand to move by air pressure. (*Clifford R. Scholes collection*)

In 1957, PTC snow sweeper C-36 is at the Southern Depot waiting for the next assignment. This single-truck snow sweeper was built in 1896 as No. 2917 by the Union Traction Company and was scrapped by PTC on February 26, 1958. (*Clifford R. Scholes collection*)

Double-truck PTC snow sweeper No. C-120 is at Southern Depot in 1957. Delivered on November 28, 1923 by the J. G. Brill Company, the snow sweeper was scrapped in October 21, 1960. (*Clifford R. Scholes collection*)

From left to right on April 27, 1968 at Luzerne Depot are double-truck snow sweeper No. C-126 delivered by J. G. Brill Company on December 26, 1923; PCC car No. 2530 (delivered in December 1940 and scrapped in 1973; and PCC car No. 2567 (delivered January 1941 and scrapped in 1973). (*Kenneth C. Springirth photograph*)

On February 22, 1969, double-truck snow sweepers C-134 (rebuilt from plow E-221 on October 21, 1925) and C-150 (rebuilt from plow E-211 on October 21, 1927) are at the Germantown Depot. There were twenty-four double-truck plows rebuilt into snow sweepers by J. G. Brill Company, because the plows were too large to effectively work in Philadelphia's narrow streets. (*Kenneth C, Springirth photograph*)

Double-truck crane car No. W-54, built by J. G. Brill Company on January 31, 1924, is at the 3rd and Courtland Shop in 1973. In 1937, two 1-ton hoists replaced the car's jib cranes, and new cabs were installed with a reduced width below the windows that made it possible to haul rail that was longer than the car by allowing the rail to overhang at either end of the car. (*Clifford R. Scholes collection*)

In 1973, drop side motor car No. W-53 is at the 3rd and Courtland Shop. Built by J. G. Brill Company on January 29, 1924, a jib crane and one cab were removed from this car in April 1959 making this car a single-end flat motor car. (*Clifford R. Scholes collection*)

On May 31, 1968, the 3rd and Courtland Shop is the location of W-56 drop side work motor car built January 31, 1924; W-40 differential dump motor car built October 18, 1923; and W-41 differential dump motor car built October 29, 1923. All of these work cars were built by J. G. Brill Company. (*Kenneth C. Springirth photograph*)

With its deck covered by a thin layer of snow, single-end, double-truck, flat car No. W-56 is at the 3rd and Courtland Shop in a lineup of cars in 1975. Along with the snow sweepers, the class W cars were durable cars. (*Clifford R. Scholes collection*)

At the 3rd and Courtland Shop on May 31, 1968, shop personnel are working on a switch for the Route 50 Rising Sun Avenue and Olney Avenue Loop. (*Kenneth C. Springirth photograph*)

Steel hopper car No. D-7 (like a regular coal car with an operating cab at each end) is ready for its next assignment at the 3rd and Courtland Shop on May 31, 1968. This double-truck car was originally No. 2624 built in 1908. (*Kenneth C. Springirth photograph*)

10

Trackless Trolleys

PRT opened the Route 80 (Oregon Avenue) trackless trolley line in Philadelphia on October 14, 1923 using nine "Rail-Less" vehicles built by J. G. Brill Company. In 1935, these were replaced by eight new trackless trolleys made by J. G. Brill Company. After the January 1, 1940 reorganization of PRT into the Philadelphia Transportation Company (PTC), fifty new trackless trolleys model 40SMT were purchased from J. G. Brill Company, replacing streetcars on the 11-mile Route 61 (Ridge Avenue) line on October 5, 1941. Ten more of these vehicles were added during 1942. This heavily used line passed through major industrial areas plus large residential neighborhoods intersecting numerous streetcar lines, the Market – Frankford Elevated Subway, Broad Street Subway, and the Ridge – Camden Spur. During 1946, Route 61 was transferred from Allegheny Depot to a rebuilt portion of the Ridge Depot.

Streetcar Route 29 (Tasker – Morris) was converted to trackless trolley operation on December 14, 1947. In 1948, Route 29 was extended from 28th Street to an off street loop at 33rd and Dickinson Streets.

Route 75 (Wyoming Avenue) became a bus line on April 18, 1948 and a trackless trolley line on April 19, 1948. While the Route 75 streetcar line turned south on 5th Street to Erie Avenue and turned west on Erie Avenue to a turnback at 17th Street, the trackless trolley line continued west on Wyoming Avenue with a short westbound jog on Roosevelt Boulevard and back to Wyoming Avenue where it crossed Broad Street providing a transfer point with the Broad Street Subway. It continued west on Wyoming Avenue to the Wayne and Windrim terminal loop adjacent to the Wayne Junction Station of the Reading Railroad.

Route 59 (Frankford to Bustleton) became a trackless trolley line on June 25, 1950. Since Route 59 was owned by the City of Philadelphia, the twenty-eight model TC-46 Marmon-Herrington trackless trolleys used on that line were owned by the city and operated by PTC. Both Routes 59 and 75 were stationed at the Frankford Depot.

Route 66 (Frankford Avenue) was converted to bus operation on July 30, 1955 and on September 11, 1955 received forty-three city-owned model TC-49 trackless trolleys built by Marmon-Herrington Company. Route 66 was long busy feeder from the Bridge Street terminal of the Market – Frankford Elevated via Frankford Avenue to the City line terminus adjacent to Bucks County. Philadelphia had three disconnected trackless trolley networks. Northeast Philadelphia Routes 59, 66, and 75 operated out of Frankford Depot. South Philadelphia Routes 29 and 79 operated out of Southern Depot. Route 61 operated from Allegheny Depot and later Ridge Depot. There was never any two wire connection between these three networks.

After PTC came under National City Lines control, for a few years no more lines were converted to trackless trolley operation. Effective August 16, 1959, Route 61 was bus operated on weekends and holidays. Route 80 was converted to bus operation on May 15, 1960 followed by Route 61 on March 13, 1961. With surplus trackless trolleys, PTC decided to convert Route 79 to trackless trolley operation, because most of its line poles were in place from its former streetcar operation. Route 79 (Snyder Avenue) was converted to trackless trolley on June 13, 1961 making it the only Philadelphia trackless trolley route created from a bus route. New residential development along Route 66 necessitated increasing its capacity, which was addressed by adding a third set of overhead wires from Cheltenham Avenue to Wellington Street in 1962. This allowed peak hour express service to pass the local service. On April 2, 1966, the eastern end of Route 75 was cutback from the Bridesburg neighborhood of Richmond Street to the Margaret – Orthodox station of the Market – Frankford Elevated Line.

SEPTA acquired PTC on September 30, 1968. The trackless trolley fleet was replaced by 110 new AM General Corporation trackless trolleys beginning in 1979. However, manufacturing defects that had to be corrected delayed their general use until the early 1980s. Ridership decline resulted in only about half of the trackless trolley fleet being used by the late 1990s. On June 9, 2002, Routes 59, 66, and 75 were temporarily converted to bus operation, because of the reconstruction of the Frankford Elevated overhead structure and the Frankford Depot garage. Buses replaced Route 29 trackless trolleys on February 23, 2003 necessitated by a detour for the rebuilding of Tasker homes near the western end of the line followed by the conversion of Route 79 to bus operation on September 7, 2003. In October 2006, SEPTA ordered thirty-eight New Flyer low-floor trackless trolleys. On April 14, 2008, Routes 66 and 75 each received one to two new trackless trolleys followed by Route 59 on April 15, 2008. Within several weeks enough new trackless trolleys were available to completely service Routes 59, 66, and 75.

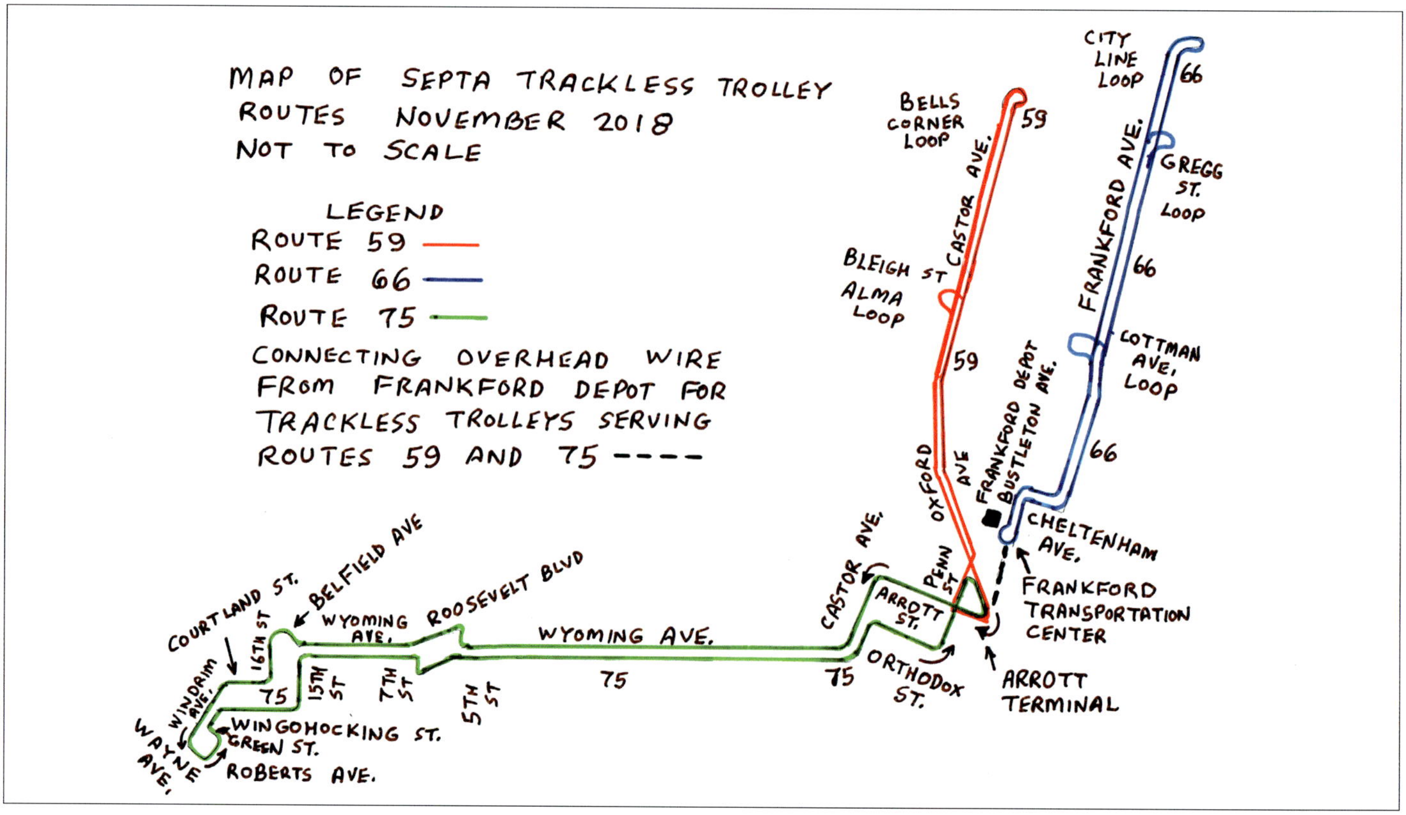

This map shows the three surviving SEPTA trackless trolley Routes 59, 66, and 75 in November 2018.

In 1979, at Frankford Depot, AM General trackless trolley model 10240T was the first of 110 new trackless trolleys (Nos 800–909) built by AM General Corporation in 1978 followed by the remainder in 1979. In the rear on the left side of the picture is Marmon-Herrington Company model TC49 trackless trolley No. 325 built in 1955 and completely rebuilt in 1978. (*Clifford R. Scholes collection*)

Eastbound Route 29 trackless trolley No. 808 (model 10240T built by AM General Corporation in 1979) is on Morris Street crossing Broad Street in the Newbold neighborhood of South Philadelphia on July 13, 1992. This crosstown route largely operated westbound on Tasker Street and eastbound on Morris Street operating between 33rd and Dickinson Street in the west and Water Street in the east. (*Kenneth C. Springirth photograph*)

On July 13, 1995, Route 29 trackless trolley No. 810 (built by AM General Corporation) is on Tasker Avenue at 12th Street in the Passyunk Square neighborhood of South Philadelphia. The route sign with a dash through '29' indicates this vehicle will short turn at 23rd and Tasker Streets, probably to go to the Southern Depot. The AM General Corporation only built 110 trackless trolleys for SEPTA and 109 for Metro Transit in Seattle, Washington. (*Kenneth C. Springirth photograph*)

Ridge Avenue at Vine Street is the location of Route 61 trackless trolley No. 236 in 1961. Seating forty-four passengers, this was one of sixty-five model TC44 trackless trolleys (Nos 201–265) built by J. G. Brill Company in 1947. (*Clifford R. Scholes collection*)

In 1961, Route 59 trackless trolley No. 487 is on Castor Avenue at Bleigh Avenue. Owned by the City of Philadelphia and seating forty-six passengers, this was one of twenty-eight model TC46 trackless trolleys (Nos 471–498) built in 1949 by Marmon-Herrington Company. (*Clifford R. Scholes collection*)

On May 28, 1992, SEPTA Route 59 trackless trolley No. 880 (built by AM General Corporation) is at the Castor Avenue and Bustleton Avenue Loop in the Rhawnhurst neighborhood of Northeast Philadelphia. Route 59 operates from the Arrott Transportation Center Station of the Market -Frankford Line to Bells Corner mainly via Oxford and Castor Avenues. (*Kenneth C. Springirth photograph*)

Avenue at Pratt Street, in the Frankford neighborhood of Northeast Philadelphia on July 2, 1999, is the location of Route 59 trackless trolley No. 876 (built by AM General Corporation). (*Kenneth C. Springirth photograph*)

Castor Avenue at Knorr Street finds Route 59 trackless trolley No. 861 (built by AM General Corporation) passing by a passenger stop with no passengers in sight on September 1, 2000. (*Kenneth C. Springirth photograph*)

On June 7, 2002, Route 59 trackless trolley No. 879 (built by AM General Corporation) is on Oxford Avenue at Frankford Avenue, which is a transfer point for the Margaret – Orthodox Station of the Market – Frankford Elevated Line. Route 59 service levels have shown a significant decline with the August 26, 2018 schedule showing a midday weekday frequency of fifteen minutes, Saturday frequency of twenty minutes, and Sunday frequency of thirty minutes compared with September 3, 1967 midday weekday and Saturday frequency of twelve minutes, plus additional short-turn service from Cottman Avenue and Sunday frequency of twelve minutes. (*Kenneth C. Springirth photograph*)

In the bright sun of June 7, 2002, Route 59 trackless trolley No. 897 is on Castor Avenue ready to cross Hartel Avenue in the Rhawnhurst neighborhood of Northeast Philadelphia. In fiscal year 2017, Route 59 had an average weekday ridership of 4,641 and required six vehicles during rush hours. (*Kenneth C. Springirth photograph*)

On October 1, 2010, Route 59 trackless trolley No. 813 (one of thirty-eight model E40LFR trackless trolleys built by New Flyer of America Inc. in 2007–2008) is at the Alma Loop (Castor and Bleigh Avenues) for the 3:05 p.m. trip that runs only on school days. The schedule effective August 26, 2018 showed a short-turn trip running only when school was in session leaving Arrott Transportation Center at 2:34 p.m., arrived at Alma Loop at 2:53 p.m., left Alma Loop at 3:03 p.m., and arrived at Arrott Transportation Center at 3:23 p.m. (*Kenneth C. Springirth photograph*)

Route 59 trackless trolley No. 801 (built by New Flyer of America Inc.) is waiting for departure time at the Arrott Transportation Center on October 1, 2010. Located at Frankford and Oxford Avenues, this was originally known as the Margaret - Orthodox Elevated Station of the Market - Frankford Line (*Kenneth C. Springirth photograph*)

Frankford Avenue at Cottman Avenue, in Northeast Philadelphia's Mayfair neighborhood, in 1980 is the location of Route 66 trackless trolley No. 883 (built by AM General Corporation). Route 66 opened as a streetcar line on October 3, 1895 on Frankford Avenue from Bridge Street to City Line. Transformed from 1920s farmland, Mayfair featured bigger row homes with front yards. Trackless trolleys took over Route 66 on September 11, 1955. (*Clifford R. Scholes collection*)

1980, Route 66 trackless trolleys No. 873 (built by AM General Corporation in 1980) and No. 325 (one of forty-three model TC49 trackless trolleys (Nos 301–343) built by Marmon-Herrington Company in 1955) are waiting for departure time at the City Line Loop. No. 325 was completely rebuilt in 1978, plus received the new SEPTA red, white, and blue paint scheme as a prototype for rebuilding Nos 301–343. This was the only trackless trolley to receive this rebuilding, because the decision was made to buy 110 trackless trolleys from AM General Corporation. (*Clifford R. Scholes collection*)

Route 66 trackless trolley No. 845 (built by AM General Corporation in 1980) is on Frankford Avenue at the Morrell Avenue passenger stop in the Morrell Park neighborhood of Northeast Philadelphia in 1983. Note the $1.058 price for regular gas and the vintage automobiles in the picture. (*Clifford R. Scholes collection*)

On March 21, 1992, Route 66 trackless trolley No. 909 (the highest number trackless trolley built by AM General Corporation for SEPTA in 1980) is on Frankford Avenue at Cheltenham Avenue in the Frankford neighborhood of Philadelphia. The stone wall of the North Cedar Hill Cemetery is on the right side of the picture. (*Kenneth C. Springirth photograph*)

Frankford Avenue near Academy Road on April 11, 1998 is the location of Route 66 trackless trolley No. 886 (built by AM General Corporation). (*Kenneth C. Springirth photograph*)

Route 66 trackless trolley No. 892 (built by AM General Corporation) is turning from Ryan Avenue to Leon Street on July 31, 1998. The red background destination sign for this Cottman Avenue short-turn trip for Route 66 indicates that it will be an express trip to the Bridge Street Terminal. (*Kenneth C. Springirth photograph*)

On April 2, 1999, Route 66 trackless trolley No. 885 (built by AM General Corporation) is leaving the Gregg Street Loop on to Frankford Avenue. Route 66 has shown a service level decline with the August 28, 2018 schedule showing no Gregg Street cutback service on Saturday and Sunday frequency of sixteen minutes compared to January 21, 1968 when there was Saturday Gregg Street cutback service and Sunday frequency was twelve minutes or less. (*Kenneth C. Springirth photograph*)

Route 66 trackless trolley No. 820 (built by New Flyer of America Inc.) is on Cheltenham Avenue turning onto Frankford Avenue passing the North Cedar Hill Cemetery on October 29, 2010. In fiscal year 2017, Route 66 had an average weekday ridership of 10,367 and required fifteen vehicles during rush hours. (*Kenneth C. Springirth photograph*)

On October 29, 2010, Route 66 trackless trolley No. 814 (built by New Flyer of America Inc.) is leaving the Gregg Street Loop with the destination sign showing Frankford Transportation Center. The August 26, 2018 schedule showed the short turn Gregg Street Loop was used by fifty-one Monday to Friday trips; however, it was not used for any regularly scheduled Saturday, Sunday, or holiday trips. (*Kenneth C. Springirth photograph*)

Street near Cottman Avenue finds Route 66 trackless trolley No. 811 (built by New Flyer of America Inc.) waiting for departure time on October 29, 2010. This is the short-turn Cottman Avenue Loop, which was used by fourteen morning and eleven afternoon rush-hour trips as shown on the SEPTA August 26, 2018 Route 66 schedule. (*Kenneth C. Springirth photograph*)

In 1972, Route 75 trackless trolley No. 263 (built by J. G. Brill Company in 1947) is on Wyoming Avenue at D Street in the Feltonville neighborhood of North Philadelphia. Route 75 connects the Arrott Transportation Center (originally Margaret - Orthodox) Station of the Market - Frankford Elevated Line mainly via Wyoming Avenue to Wayne Junction for commuter train connections. (*Clifford R. Scholes collection*)

Route 75 trackless trolley No. 489 (built by Marmon-Herrington Company) is turning from Orthodox Street to Penn Street in the Frankford neighborhood of Northeast Philadelphia in 1976. (*Clifford R. Scholes collection*)

In 1976, Wyoming and Rising Sun Avenues is the location of eastbound Route 75 trackless trolley No. 487 (built by Marmon-Herrington Company). Behind is a Route 50 streetcar. Route 75 has shown a significant service level decline with the August 26, 2018 schedule showing a midday weekday frequency of twenty minutes and Saturday–Sunday frequency of twenty-seven to thirty minutes compared to September 3, 1967, showing a midday weekday frequency of twelve minutes or less and a Saturday frequency of twelve to fourteen minutes and a Sunday frequency of fourteen to fifteen minutes. (*Clifford R. Scholes collection*)

Windrim Avenue at Germantown Avenue (with the Route 23 streetcar tracks in the street) is the location of Route 75 trackless trolley No. 325 in 1980. This was the only Marmon-Herrington Company-built trackless trolley that was completely rebuilt in 1978 as well as repainted in the new SEPTA red, white, and blue paint scheme. SEPTA made the decision to purchase 110 new trackless trolleys from AM General Corporation in a joint order with Metro Transit in Seattle, Washington, that received 109 new trackless trolleys. (*Clifford R. Scholes collection*)

On May 28, 1992, Route 75 trackless trolley No. 894 (built by AM General Corporation) is at the Margaret - Orthodox (later renamed Arrott Transportation Center) Station of the Market - Frankford Elevated Line. Originally Route 75 continued east to Bridesburg until it was cut back to the Margaret - Orthodox station on April 3, 1966. (*Kenneth C. Springirth photograph*)

Windrim Ave at Greene Street on October 19, 2008 is the location of a photo stop for a Friends of Philadelphia excursion with trackless trolley No. 837 (built by New Flyer of America Inc.) on Route 75. (*Kenneth C. Springirth photograph*)

Route 75 trackless trolley No. 820 is coming off Frankford Avenue and entering the Arrott Transportation Center (originally Margaret - Orthodox) Station under the Market - Frankford Elevated on February 5, 2010. In fiscal year 2017, Route 75 had an average weekday ridership of 3,487 and required eight vehicles during rush hours. (*Kenneth C. Springirth photograph*)

On March 3, 1969, Route 79 trackless trolley No. 256 (built by J. G. Brill Company) is westbound on Snyder Avenue crossing Broad Street. Streetcars operated on Route 79 from January 1, 1912 to replacement by buses on October 28, 1956. Route 79 was converted to trackless trolley operation on June 18, 1961 operating on Snyder Avenue from West 29th Street to Dilworth Street. (*Kenneth C. Springirth photograph*)

Route 79 trackless trolley No. 205 (built by J. G. Brill Company) is on Snyder Avenue at 11th Streets on July 24, 1978. The trackage on Snyder Avenue was a cutback for streetcar Route 23. (*Kenneth C. Springirth photograph*)

On November 24, 2000, Route 79 trackless trolley No. 842 (built by AM General Corporation) is westbound on Snyder Avenue ready to cross 12th Street. Route 23 streetcars used Snyder Avenue between 12th and 11th Streets as a cutback until replaced by buses on February 21, 1992. (*Kenneth C. Springirth photograph*)

At the western terminus of Route 79 on Snyder Avenue at 29th Street, trackless trolleys Nos 897 and 823 (both built by AM General Corporation) are waiting for departure time on April 18, 2003. On September 7, 2003, Route 79 was converted to bus operation. (*Kenneth C. Springirth photograph*)

11

Fairmount Park Trolleys

The December 15, 1894 notice in the *Philadelphia Times* newspaper announced that on December 14 in Trenton, NJ, "Articles of incorporation of the Fairmount Park Transportation Company were filed in the office of the Secretary of State today. The corporation is organized for the purpose of building railways in Fairmount Park, Philadelphia. The total capital is $2,000,000." A portion of the line opened on November 10, 1896. During 1896, J.G. Brill Company built ten closed motor cars, twenty motorized open cars, and thirty open trailer cars followed by ten more closed cars in 1897. The completion of the Strawberry Mansion Bridge over the Schuylkill River resulted in the entire line opening on June 13, 1897. On the 8.83-mile-long line, track gauge was 5 feet 2.5 inches compared to 5 feet 2.25 inches on Philadelphia Rapid Transit Company. The line had seventeen stations, of which fifteen were in Fairmount Park. Woodside Park was opened in 1897 by the trolley company on the south side of Ford Road just east of its intersection with Monument Road and Conshohocken Avenue. The purpose of Woodside Park with its Dentzel Carousel, roller coasters, and lake was to increase trolley car ridership. On December 11, 1918, facing financial problems, the company was sold to the Fairmount Park Transit Company. With declining revenues, the company filed for bankruptcy 1935. The Pennsylvania Public Utility Commission granted permission to abandon the line on May 17, 1946, and service ended on September 9, 1946. Woodside Park closed in 1955.

The *Philadelphia Inquirer* of November 25, 2018 in an article "A Hike Into the Past" noted that the Fairmount Park Conservancy is working with the Philadelphia Parks and Recreation to turn the former trolley right of way into a 4.5-mile Trolley Trail. A portion of that trail has been completed, and the goal is to create a continuous trail of that scenic route.

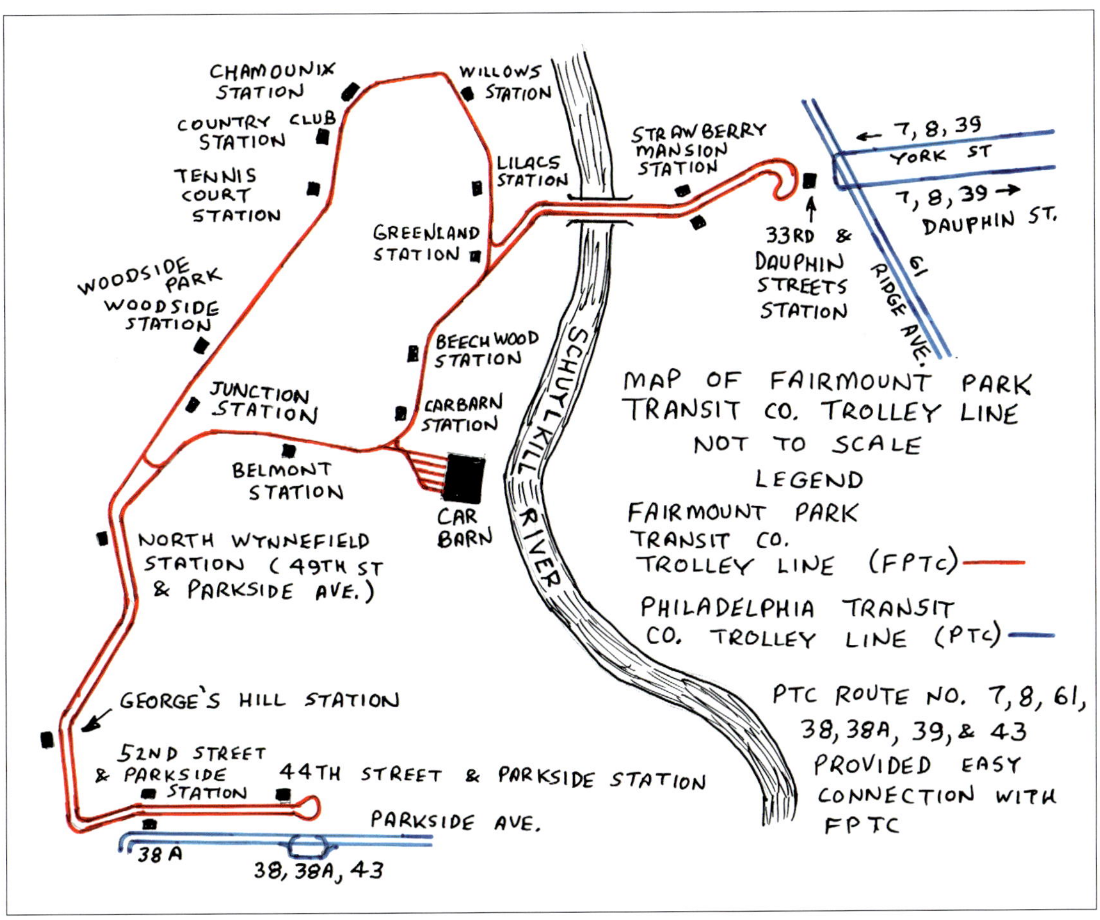

This map of the Fairmount Park Transit Company trolley line shows the line that was opened in 1896 and closed in 1946 plus its PTC streetcar connections.

In 1937, Fairmount Park Transit Company car No. 15 is on the siding at Woodside Station. There were twenty 20 open motorized cars (Nos 11–12 and 14–31) without any No. 13 because of superstition. These twelve bench cars were built by J. G. Brill Company in 1896. The trolley line was important to Fairmount Park making it easier to get around the park plus transporting riders to Woodside Park for picnics, fireworks shows, and concerts. (*Clifford R. Scholes collection*)

Fairmount Park Transit Company open trailer car No. 55 is at the 44th and Parkside Station with its Red Terra Cotta style roof in 1942. In 1896, J. G. Brill Company built these thirty open-trailer cars (Nos 32–61). Financial problems resulted in service ending on September 9, 1946. After track, bridges, station buildings, and remaining infrastructure were sold at the November 6, 1946 auction, they were promptly removed with nature reclaiming the area. In 2019, remains of the trolley line, including classy stone bridges are coming to life along the 4.5-mile Trolley Trail being completed by the Fairmount Park Conservancy. (*Clifford R. Scholes collection*)

12

Penn's Landing Trolley

A heritage streetcar (trolley) line was established along a 1.1-mile section of no longer used former Philadelphia Belt Line Railroad trackage that was owned by Consolidated Rail Corporation from the Benjamin Franklin Bridge to Fitzwater Street (Pier 51) in the center private right of way median of Delaware Avenue. The goal was to attract tourists and enhance redevelopment of the area. Grants from the City of Philadelphia and a local bank paid for the installation of poles and overhead wire. Volunteers from the Buckingham Valley Trolley Association (BVTA) began operating historic trolley cars on September 5, 1982. BVTA was a nonprofit group that purchased vintage trolley cars and preserved them for future use. The trolleys were stored in a building on city-owned Pier 51. BVTA had operated trolleys on the New Hope & Ivyland Railroad in Bucks County in the 1970s until the railroad ownership changed, and the trolley operation ended. The Penn's Landing Trolley had a two-man crew with a motorman operating the trolley and the conductor collecting tickets. Round trip fare was $1 for adults and 50 cents for children. Ridership peaked in 1987 when the signing of the United States Constitution brought tourists to the waterfront. Since the Penn's Landing Corporation needed pier space for development, the line made its last run on December 17, 1995. The trolley wire and poles were removed by March 1996. Many of the trolleys were moved to the Electric City Trolley Museum at Scranton, Pennsylvania.

On November 7, 1982, former Philadelphia Suburban Transportation Company (PSTC) car No. 26 is operating on Delaware Avenue near the Delaware River in Center City Philadelphia. It was Philadelphia Transportation Company (PTC) car No. 4045 delivered in July 1918 and sold to PSTC in October 1942 where it was renumbered 21 and later 26. This was originally part of an order of 100 cars (Nos 4000–4099) built by J. G. Brill Company in 1918 for the Philadelphia Rapid Transit Company, which became PTC on January 1, 1940. (*Kenneth C. Springirth photograph*)

In April 1987, former PSTC car No. 80 is on Delaware Avenue with the Benjamin Franklin Bridge in the background. This was one of ten 80 series cars built by J. G. Brill Company in 1932. Powered by four General Electric type 301B motors, each car weighed 41,980 pounds and seated sixty-one passengers. It was acquired by the Electric City Trolley Museum in 1999. (*Kenneth C. Springirth photograph*)

Delaware Avenue, south of Market Street, is the location of former PSTC car No. 26 and behind it is PSTC car No. 80 in April 1987. (*Kenneth C. Springirth photograph*)